Anatomy of Flowering Plants

An introduction to structure and development

Paula Rudall PhD FLS

Jodrell Laboratory
Royal Botanic Gardens
Kew

Edward Arnold

First published in Great Britain 1987 by
Edward Arnold (Publishers) Ltd,
41 Bedford Square, London WC1B 3DQ

Edward Arnold (Australia) Pty Ltd,
80 Waverley Road, Caulfield East,
Victoria 3145, Australia

Edward Arnold, 3 East Read Street,
Baltimore, Maryland 21202, USA

British Library Cataloguing in Publication Data

Rudall, Paula
 Anatomy of flowering plants: an introduction to structure and development.
 1. Angiosperms—Anatomy
 I. Title
 582.13'044   QK641

 ISBN 0-7131-2950-6

Text set in 10/11 Plantin Compugraphic
by Colset Private Limited, Singapore
Printed and bound in Great Britain by
Whitstable Litho, Whitstable, Kent

Contents

Preface

Plant anatomy is a fascinating subject, now unfortunately often neglected in University courses in the United Kingdom, and sometimes regarded by students as 'old-fashioned', probably because of the wealth of reference works from the last century, and the simplicity of many of the basic techniques, in an age of complex technology. However this apparent simplicity can be misleading; the interpretation of prepared material requires some knowledge and experience. Often the contemporary plant anatomist has recourse to modern equipment such as the scanning or transmission electron microscopes, or refined methods such as the use of cinematography to analyse serial sections. Although a great deal of excellent work in plant anatomy was undertaken in the nineteenth and early twentieth centuries, particularly in Germany, this century has seen an enormous progression of research in many countries of the world, such as the United States, India, Germany, Holland and the United Kingdom. At the present time much active research ensures that plant anatomy is still a lively field. This book sets out to present a basic introduction to the subject to students of botany and other related disciplines. Different aspects of development are discussed and some emphasis is given to areas of currently active research, often with controversial interpretations, to try to capture the interest of the potential researcher. It is impossible to give a fully comprehensive account of such a broad subject in a book of this size, and examples of recommended further general reading are given on page 71.

I am very grateful to Mary Gregory of the Jodrell Laboratory for critically reading the manuscript, and to David Cutler and other staff and visitors of the Anatomy section, Jodrell Laboratory, whose help in the past has contributed to the writing of this book. Thanks are also due to the Director of the Royal Botanic Gardens, Kew, and the Keeper of the Jodrell Laboratory for permission to use facilities.

1

General Plant Structure

At maturity higher plants consist of several organs, which in their turn are made up of tissues. Tissues are composed of groups of similar individual cells and may be interspersed with other cell types (idioblastic cells), which differ from those of the surrounding tissue.

1.1 Organization

The concept of plant organs has long been investigated and discussed by plant morphologists (e.g. Arber, 1950). Three main organs are widely recognized: root, stem and leaf (Fig. 1.1). Roots are usually found in the soil, from which they extract moisture and nutrients, although there are many examples of plants with aerial roots. Most dicotyledons have a tap root with side branches (lateral roots), which develops from the seedling radicle. However in monocotyledons the seedling radicle commonly dies at an early stage and the roots of the mature plant are borne on the stem and may be branched or unbranched (adventitious roots). The stem and leaves, which together comprise the shoot (Arber, 1950), typically occur above ground level, although some stems are modified into underground perennating or storage organs such as corms or rhizomes, and underground bulbs have specialized leaves or leaf bases. Stems are divided into nodes and internodes, the node being the point of attachment of the leaf or leaves. Each leaf subtends an axillary bud. The pattern of arrangement of the leaves on the stem is called phyllotaxis. At the onset of flowering the shoot apical meristem changes from a vegetative to a reproductive apex and subsequently produces flowers (chapter 5). Flowers are classically interpreted as modified shoots, and the floral parts (sepals, petals, stamens and carpels) as modified leaves borne on a central axis (Esau, 1965).

1.2 Cell structure

Plant cells typically have a cell wall containing a living protoplast (Fig. 1.2). The layer between the cell walls of neighbouring cells is termed the middle lamella. Many cells, when they have ceased growth, develop a secondary cell

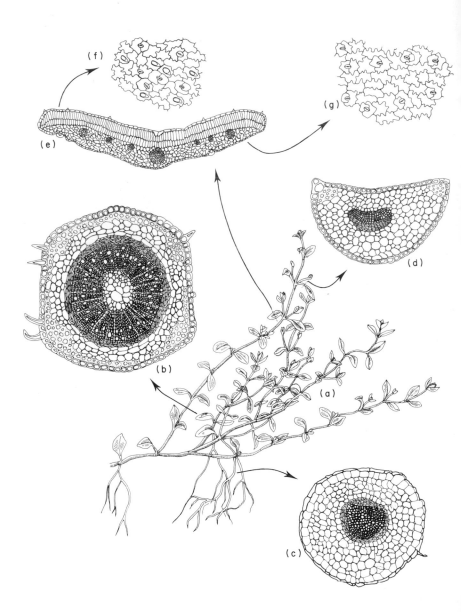

Fig. 1.1 Whole plant **(a)** of *Thymus* sp. (thyme) with cross sections of vegetative organs; **(b)** stem, **(c)** root, **(d)** petiole, **(e)** leaf, **(f)** adaxial leaf surface, **(g)** abaxial leaf surface.

wall which is deposited on the inside surface of the primary wall. Both primary and secondary walls consist of cellulose microfibrils embedded in a matrix and oriented in different directions. Secondary walls are mostly cellulose, but primary walls commonly contain a high proportion of hemi-cellulose in the gel-like matrix, affording a greater degree of plasticity to the wall of the growing cell. The secondary wall may also contain deposits of lignin (e.g. in sclerenchyma) or suberin (e.g. in many periderm cells), and often appears lamellated. Thin areas of the primary wall, which usually correspond with thin areas of the wall of neighbouring cells, are primary pit fields, and usually have protoplasmic strands (plasmodesmata) passing through them and connecting the protoplasts of adjacent cells. The connected living protoplasts are sometimes called the symplast. Primary pit fields often remain as thin areas of the wall even after a secondary wall has been deposited, and are then termed pits, or pit-pairs if there are two pits connecting adjacent cells. Pits may be simple, as in most parenchyma cells, or bordered, as in tracheary elements. In simple pits the pit cavity is of more or less uniform width, whereas in bordered pits the secondary wall arches over the pit cavity so that the opening to the cavity is narrow. Through a light microscope the outer rim of the primary pit field appears as a border around the pit opening.

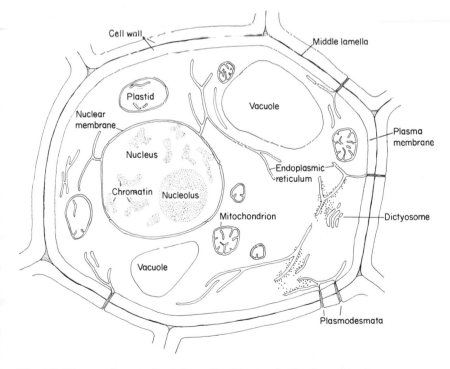

Fig. 1.2 Diagram of a generalized plant cell to illustrate details of protoplasmic contents.

The cell protoplast is contained within a plasma membrane. It consists of cytoplasm enclosing bodies such as the nucleus, plastids and mitochondria, and also non-protoplasmic contents such as oil droplets, starch granules and crystals. The nucleus, which is bounded by a nuclear membrane, often contains one or more recognizable bodies (nucleoli) together with the chromatin in the nuclear sap. During cell division the chromatin becomes organized into chromosomes. Most cells have a single nucleus, but examples of multinucleate cells include the non-articulated laticifers found in many plant families, such as Euphorbiaceae (Fig. 2.3). These laticifers are coenocytes, or groups of cells that have undergone division without corresponding wall formation.

Mitochondria and plastids are also surrounded by membranes. Plastids are larger than mitochondria, and are classified into different types depending on their specialized function. For example, chloroplasts are plastids that contain chlorophyll, for photosynthesis. They occur in all green cells, but are most abundant in the leaf mesophyll, the primary photosynthetic tissue (section 4.3). Membranes occur widely throughout the cytoplasm, sometimes bounding a series of cavities, the endoplasmic reticulum, or dictyosomes, which are associated with secretory activity. Vacuoles are cavities in the cytoplasm, usually colourless and containing a watery sap. They vary in size and shape in the different cell types, and often vary considerably during the life of a cell. (See also Burgess, 1985.)

Many cells have non-protoplasmic inclusions such as oils (which usually appear as droplets), dark brown tannins, and starch granules and calcium oxalate crystals, both of which take many forms. Starch is deposited particularly in storage tissues, and the granules, which often appear layered due to the successive deposition of concentric rings, are formed in plastids (amyloplasts). Starch granules sometimes have characteristic shapes in different plants. Calcium oxalate crystals may be prismatic or compound, and solitary or in groups. Styloids are long prismatic crystals; raphides are thin needle-like crystals, often in aggregates; druses (cluster crystals) are compound structures; and crystal sand is a group of many minute separate crystals in one cell (Franceschi and Horner, 1980). A few plant families, such as Gramineae, Cyperaceae or Palmae, sometimes have characteristic silica bodies contained in certain cells, such as idioblastic epidermal cells.

1.3   Tissue distribution and origin

Mature plant tissues can broadly be divided into three main groups: epidermis, ground tissue and vascular tissue, each distributed throughout the plant body and often continuous between the various organs.

The epidermis is the outermost layer of cells, covering the entire plant surface, and usually uniseriate. It is a primary tissue derived from the outermost layers of the shoot and root apical meristems, although anticlinal divisions (at right angles to the surface) may occur in mature epidermal cells in step with stem or root thickening. In older stems and roots the epidermis often splits and peels off following an increase in thickness, and is replaced by a periderm (section 2.4; Fig. 2.13). In some monocotyledon roots the epidermis is worn away by the soil, and is replaced by an exodermis, formed by cell wall thickening in the outer cortical layers. The epidermis includes many specialized cell types such as root hairs (section 3.2.1), stomata and trichomes on stems and leaves (section 4.2), and secretory tissues such as nectaries on various above-ground organs, particularly flowers (section 5.6).

Ground tissue often has a mechanical function, or it may be concerned with storage or photosynthesis. In general it consists of parenchyma, sclerenchyma or collenchyma (Fig. 1.3), often interspersed with idioblasts and secretory cells or canals. It forms the bulk of primary plant tissues and occupies the areas within the epidermis that are not vascular tissue or cavities. Ground tissue is initially formed at the apical meristems but may be supplemented by intercalary growth, and in monocotyledons by tissues differentiated from primary and secondary thickening meristems. In dicotyledons the ground tissue of secondary xylem (wood), formed by the vascular cambium, is supplied by fibres and axial parenchyma (section 2.3.1). The central area of ground tissue in older stems and thick leaves often breaks down to form a central cavity (Fig. 2.2).

Vascular tissue consists of xylem and phloem, and may be primary or secondary in origin. Primary vascular tissue is derived from procambium, itself produced by the apical meristems, and also by the primary thickening meristem in stems of certain monocotyledons (section 2.3.2). Secondary vascular tissue is derived from the vascular cambium in dicotyledons, and from the secondary thickening meristem in monocotyledons although secondary growth is rare in monocotyledons. Both xylem and phloem are complex tissues, composed of many different cell types. Xylem is primarily concerned with water transport and phloem with food transport. Distribution of vascular tissue varies considerably between different organs and taxa.

1.4 Meristems

Meristematic tissue consists of thin walled, tightly packed cells which undergo frequent divisions. Most of the plant body is differentiated at the meristems in well-defined zones, although cells in other regions may also occasionally divide.

Apical meristems are located at the shoot apex (Fig. 2.1), where primary stem, leaves and flowers differentiate, and at the root apex (Figs 3.1, 3.2), where primary root tissue is produced. Subsequent elongation of the shoot axis may occur by random cell divisions and growth throughout the youngest internodes. This region of diffuse cell division is termed an uninterrupted meristem, and is continuous with the apical meristem. However in some plant stems, particularly in the Gramineae (grasses), most cell divisions contributing to stem elongation occur in a limited area, usually at the base of the internode, which is then called an intercalary meristem

(Fisher and French, 1976, 1978). Both intercalary and uninterrupted meristems represent growth in regions of already differentiated tissues.

Lateral meristems are located parallel to the long axis of a shoot or root, most commonly in the pericyclic region, at the junction between vascular tissue and cortex. Examples of lateral meristems are the primary and secondary thickening meristems of monocotyledons, which produce both ground tissue and vascular tissue (section 2.3.2; Figs 2.11, 2.12), and the vascular cambium of dicotyledons, which produces secondary xylem and phloem (section 2.3.1; Figs 2.6, 2.7, 2.8). Adventitious roots are often formed in the pericyclic region of stems (Fig. 3.6), and lateral roots in the root pericycle; in these cases the pericycle could be termed a lateral meristem. The phellogen or cork cambium is also a lateral meristem in the stem or root cortex, where it forms a protective corky layer when the epidermis peels away (section 2.4; Fig. 2.13). However a phellogen may also develop in the region of a wound, or at the point of leaf abscission.

1.5   Tissue and cell types

Parenchyma cells are usually thin walled and often polyhedral or otherwise variously shaped, sometimes lobed. They are the least specialized cells of the mature plant body as they most closely resemble meristematic cells. In many cases they retain the ability to divide at maturity, so are instrumental in wound healing. Parenchyma cells may occur in ground tissue or vascular tissue. Relatively specialized parenchyma cells include certain secretory cells, cells containing tannin or crystals, and chlorenchyma cells, which contain chloroplasts for photosynthesis. Parenchymatous cells may be tightly packed or with many intercellular air spaces between them. Aerenchyma is a specialized parenchymatous tissue that often occurs in aquatic plants, with a regular, well-developed system of large intercellular air spaces. Cells with living contents that do not fit readily into other categories are usually termed parenchyma cells.

Collenchyma consists of groups of axially elongated, tightly packed cells with unevenly thickened walls. This tissue has a strengthening function and often occurs in the angles of young stems (e.g. in *Thymus*, Fig. 1.1), or in the midribs of leaves, always in the ground tissue. Collenchyma cells differ from

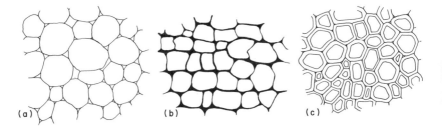

Fig. 1.3   Tissue types in cross section: (**a**) parenchyma, (**b**) collenchyma, (**c**) sclerenchyma.

fibres in that they often retain their contents at maturity and do not have lignified walls, although they may later become lignified and form fibres.

Sclerenchyma, also a supporting or protective tissue, consists of cells which have thickened, often lignified, walls, and usually lack contents at maturity. Sclerenchyma cells may occur in ground tissue or vascular tissue, either in groups or individually as idioblasts within other tissue types. They are defined as either fibres or sclereids, although there are transitional forms. Fibres are long narrow cells, elongated along the long axis of the organ concerned, and most commonly occurring in groups. Sclereids may be variously shaped and have been categorized into different types (Tschirch, 1889; Metcalfe and Chalk, 1979). Brachysclereids or stone cells are more or less isodiametric and may be present in any part of the plant. They include cells interspersed among parenchymatous tissues that develop thick secondary walls as the plant ages. Macrosclereids are rod-like cells often found in the seeds of Leguminosae. Astrosclereids are highly branched and sometimes star-shaped, and osteosclereids are bone-shaped; both types are idioblasts found most commonly in leaves. During the development of the leaf they grow intrusively into surrounding intercellular air spaces or along middle lamellae, and their shapes are to some extent dictated by the nature of the surrounding tissues.

Vascular tissue is complex and consists of several cell types: conducting cells, parenchyma cells, fibres and sometimes also sclereids, particularly in secondary phloem.

Xylem is the water conducting tissue. The conducting cells of the xylem are called tracheary elements, and may form axially linked chains (vessels).

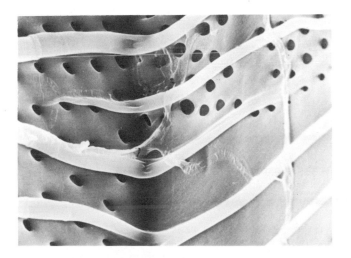

Fig. 1.4 *Tilia cordata* (lime). Inside surface of vessel element showing wall thickenings and intervascular pitting. (SEM, ×1800).

Fig. 1.5 *Betula pendula* (birch). Scalariform perforation plate. (SEM, ×320).

They usually have thickened lignified walls and lack contents at maturity. Two basic types of tracheary element can be recognized: tracheids and vessel elements. Vessel elements have large perforations in their end walls adjoining other vessel elements, whereas tracheids do not. The perforations may have one opening (simple perforation plate) or several openings, either divided by a series of parallel bars (scalariform perforation plate: Fig. 1.5), or by a reticulate mesh (reticulate perforation plate). Both tracheids and vessel elements may have bordered pits, and also wall thickenings (Fig. 1.4). Wall thickenings may be annular (in a series of rings), helical (spiral) or arranged in a scalariform or reticulate mesh. Annular and helical thickenings are the types most commonly found in the first-formed (protoxylem) elements. The last-formed primary tracheary elements (metaxylem) and also the elements of the secondary xylem typically have bordered pits in their walls. These vary considerably in size, shape and arrangement. They may be oval, polygonal or elongated (scalariform), and organized in transverse rows (opposite pitting) or in tightly packed arrangement (alternate pitting). An evolutionary series from tracheids to vessel elements is widely recognized due to the preponderance of tracheids in lower vascular plants and vessel elements in angiosperms. Similarly, scalariform pitting is considered to be the most primitive type in angiosperms, and many trends of specialization are widely established (Metcalfe and Chalk, 1983).

Phloem is the food conducting tissue. Phloem conducting cells are called sieve elements and are linked axially to form sieve tubes. Sieve elements have thin walls with characteristic sieve areas, which consist of groups of pores and associated callose. Sieve areas connect adjacent sieve elements.

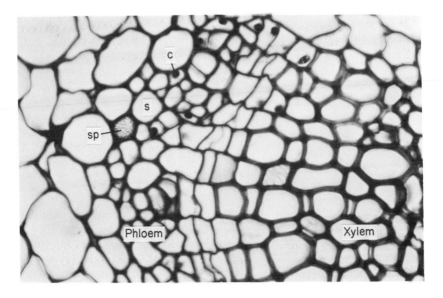

Fig. 1.6 *Vicia faba* (broad bean). Cross section of part of stem vascular tissue showing phloem and edge of xylem; s = sieve tube, c = companion cell, sp = sieve plate. (×550)

Sieve elements may be either sieve cells (most commonly found in lower vascular plants), or sieve-tube elements, which are more specialized cells found in most angiosperms. In sieve cells the sieve areas are distributed throughout the cell wall, but in sieve-tube elements they are mainly localized on the adjoining end walls, in the form of sieve plates (Fig. 1.6). Sieve plates may be simple or compound. Mature sieve elements retain their contents at maturity, but usually lack nuclei. Behnke (1972, 1975, 1981) has shown that differences in the ultrastructure of sieve tube plastids have important systematic and evolutionary applications. Sieve-tube elements commonly have associated specialized parenchyma cells, termed companion cells. These are derived from the same initial in the vascular cambium as the adjacent sieve-tube element, and perform a physiologically related function.

2

The stem

2.1 Shoot apex

Many authors have attempted to define conveniently the recognizable zones of activity which partition the shoot apical meristem of angiosperms. The most widely accepted theory is still that proposed by Schmidt (1924), which divided the central apical zone into two main regions: tunica and corpus. The tunica, which varies considerably in thickness, represents the outer (usually 1–6) layers of cells, with cell divisions mainly anticlinal (at right angles to the surface). The corpus is the region proximal to the tunica, with cell divisions orientated in all directions. The difference between the two layers is largely quantitative, and there is often slight intergradation between them, the outermost corpus layers having more anticlinal cell divisions than the layers within them. There may be variation in size and distinctness of zonation even within different stages of development of the same plant. As Clowes (1961) pointed out, a consideration of the mechanical aspects of apical growth leads to an expectation of a system in which the outer layers contribute to surface growth and the inner layers to an increase in volume.

The central apical cells of both tunica and corpus layers are sometimes relatively larger and more highly vacuolated than those on either side, and are termed tunica or corpus initials. The central region underlying the

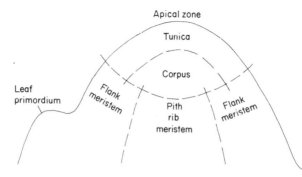

Fig. 2.1 Diagram of angiosperm shoot apex organization. (Adapted from Gifford and Corson, 1971.)

corpus layer is a rib meristem giving rise to files of cells which later become the pith. This central region is surrounded by a peripheral flank meristem which produces the procambium, cortical region, and leaf primordia (Fig. 2.1).

The vegetative shoot apex contributes to extension growth of the shoot and initiates leaf primordia. At the onset of flowering, the vegetative shoot apex changes to a reproductive apex which subsequently produces floral parts. The reproductive apex differs in size and shape and areas of mitotic activity. In general, at floral transition there is an overall increase in mitotic activity at the apex, but a proportionally greater increase among the axial apical cells than the peripheral cells (Gifford and Corson, 1971).

2.2   Primary stem structure

The plant stem is generally cylindrical, or less commonly ridged or quad-rangular (Figs 1.1, 2.2, 2.5). Primary vascular tissue usually consists of either a complete cylinder or a system of discrete vascular bundles. The cortex is the region of ground tissue between the vascular tissue and the epidermis or periderm, and the junction between the cortex and vascular tissue is the pericyclic region, from which endogenous adventitious roots may arise (section 3.3). Where there is a distinct central region of ground tissue, this is termed the pith, although in many stems the pith breaks down to form a central hollow cavity.

The stem epidermis, usually a single layer of cells, may often have stomata and trichomes, as in the leaf epidermis (section 4.2). The stem ground tissue is basically parenchymatous but may be modified into various tissue types or interspersed with fibres and sclereids. Parenchyma cells often become ligni-fied as the plant ages. Square or angled stems often have strengthening collenchyma at the angles, immediately within the epidermis. Many stems

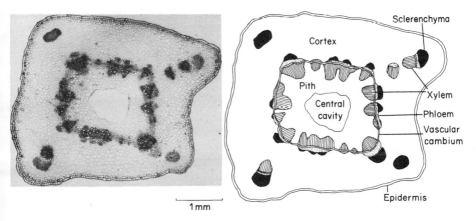

Fig. 2.2   *Vicia faba* (broad bean). Cross section of stem.

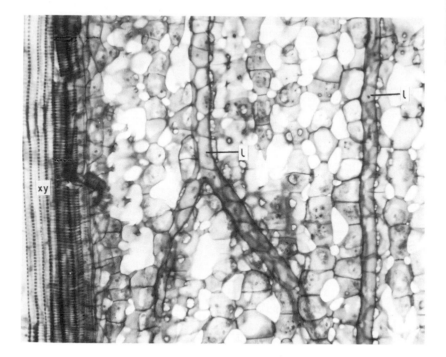

Fig. 2.3 *Euphorbia balsamifera* stem, longitudinal section of pith, showing laticifers (l) interspersed in ground parenchyma; xy = xylem. (×150).

are photosynthetic organs with a chlorenchymatous cortex, particularly in leafless (apophyllous) plants, such as many Juncaceae. Leafless plants often occur in dry, or xeric, environments, and frequently have xeromorphic modifications such as sunken stomata or large areas of sclerenchyma (Böcher and Lyshede, 1968, 1972) (see also section 4.5). Some plant stems have secretory cells or ducts in the ground tissue. For example many species of *Euphorbia* have branched networks of laticifers (latex-secreting cells) in the cortex, which often extend throughout the ground tissue of the stem and leaves (Fig. 2.3).

Plants with succulent stems, such as many Cactaceae, typically have areas of large thin-walled cells which contain a high proportion of water. Many stems store food reserves in the form of starch granules, most commonly in the inner cortex, but also frequently throughout the rest of the ground tissue. Starch is most abundant in stems that are specialized as storage or perennating organs, such as corms of *Crocus* and other seasonally active plants. Sometimes the layer of cortical cells immediately adjacent to the vascular tissue is distinct from the rest of the cortex, and packed with starch granules. It is termed a starch sheath, or sometimes an endodermoid layer or endodermis, although the component cells usually lack the Casparian thickenings typically found in the root endodermis (section 3.2.2).

2.2.1 Primary vascular system

The primary vasculature of the stem derives from the procambium near the shoot apex. Vascular bundles may be collateral, with xylem and phloem adjacent to each other, bicollateral, with phloem on both sides of the xylem, or amphivasal, with xylem surrounding the phloem.

In dicotyledons the vasculature of the internodal regions of the stem is typically arranged either in a continuous cylinder, or in a cylinder of separate or fused collateral bundles, with the phloem external to the xylem. In some stems the bundles may be bicollateral; for example in species of *Cucurbita* internal phloem is present in addition to the external phloem. Amphivasal bundles are relatively unusual in dicotyledon stems. The vascular cambium, which produces secondary vascular tissue, is normally situated between the xylem and phloem, usually in a complete cylinder. Some stems also have cortical or medullary (pith) bundles, which may be associated with the leaf vasculature.

The nodal anatomy of dicotyledons is often characteristic of families and therefore taxonomically useful (Howard, 1979), particularly the number and arrangement of leaf traces and gaps. Leaf and stem vasculature are connected at the node, with gaps in the stem vascular cylinder beneath their point of contact (Fig. 2.4). Nodes may be unilacunar, trilacunar or multilacunar, depending on the number of gaps in the main vascular cylinder. This feature is clearer in stems where the vascular cylinder is complete, especially where a limited amount of secondary thickening has taken place, with the result that nodal anatomy has been studied far more extensively in woody than herbaceous plants (Howard, 1974, 1979). The number of leaf traces departing from each gap is also variable; for example in *Clerodendron* two traces depart from a single gap, and in *Prunus* a single trace departs from each of three gaps in the central vascular cylinder. Nodal vasculature is further complicated by the axillary bud traces, which are connected to the main stem vasculature immediately above the leaf gaps. In the majority of cases there are two traces to each bud or branch. In an examination of how tree branches are attached to trunks, Shigo (1985) showed that our understanding of this is often imperfect, and it is important in a knowledge of the spread of tree diseases, and in pruning. The vascular tissue at the base of the branch is orientated abruptly downwards and forms a 'collar' around the branch base. This branch collar is enveloped by a trunk collar, which links the vascular tissue of the trunk above and below the branch. There is no direct connection of xylem from the trunk above a branch into the branch xylem, as the tissues are oriented at right angles to each other. If a branch dies, a protection zone forms around its base to prevent a spread of infection into the trunk, and the branch is often shed.

In cross sections of monocotyledon stems (Fig. 2.5) the vascular bundles are scattered apparently randomly throughout the central ground tissue, or

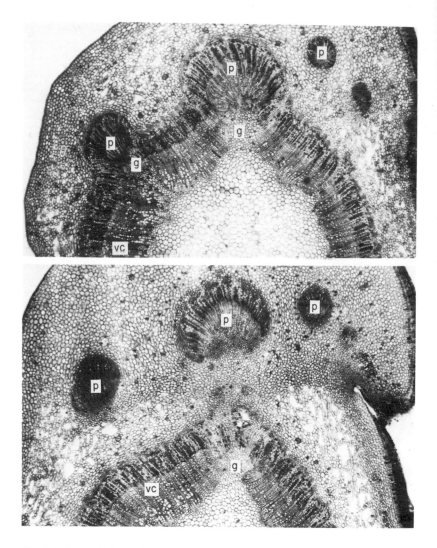

Fig. 2.4 *Prunus lusitanica* (laurel). Cross sections of part of stem at node, showing departure of petiole vasculature (p) from main vascular cylinder (vc), leaving leaf gaps (g). (×35).

sometimes arranged in two or more distinct rings. Bundles may be collateral, bicollateral, or most commonly amphivasal (especially in rhizomes), and lack a vascular cambium. Cortex and pith are frequently indistinct, although the cortex may sometimes be defined by an endodermoid layer, or a distinct ring of vascular bundles, or in some stems, particularly inflorescence axes, by a cylinder of sclerenchyma enclosing the majority of vascular bundles. In a series of investigations using cinematographic techniques of

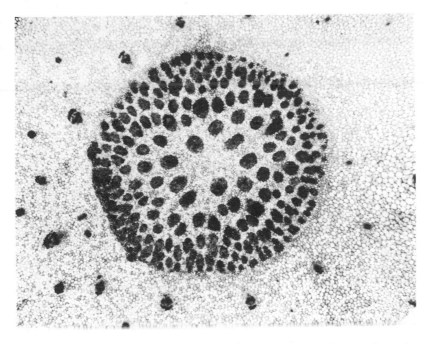

Fig. 2.5 *Libertia paniculata.* Cross section of inflorescence axis, showing central vascular region with numerous bundles and outer cortex with few bundles. (× 15).

analysis, Zimmermann and Tomlinson (1965, 1966, 1972) showed that the vascular system of monocotyledons is often extremely complex. Each major bundle, when traced on an upward course from any point in the stem, may have several bridges or branches before passing into a leaf. One of its major branches then continues a similar upward course towards the apex. In some palms there may be literally thousands of vascular bundles in a single cross section of the stem, although in most other monocotyledons the number is much lower.

2.3 Stem growth in thickness

Increase in height, achieved by growth at the apical meristem, is often accompanied by a corresponding increase in stem thickness. This is brought about by means of different meristems in dicotyledons and monocotyledons.

2.3.1 Secondary thickening in dicotyledons
In dicotyledons secondary vascular tissue (both xylem and phloem) is produced by the vascular cambium, which usually becomes active at a short distance behind the stem apex. The amount of secondary vascular tissue

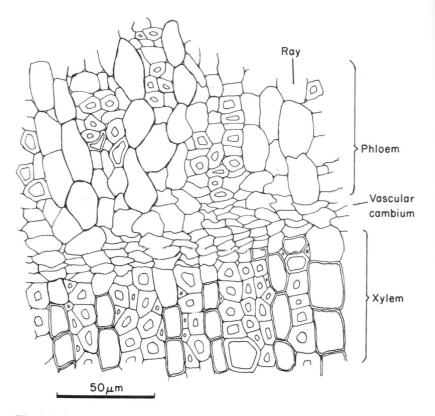

Fig. 2.6   *Prunus lusitanica* (laurel). Cross section of stem in region of vascular cambium.

produced is extremely variable, depending on the habit of the plant; in a few herbaceous dicotyledons (e.g. species of *Ranunculus*) it is completely absent. The vascular cambium generates secondary xylem (wood) at its inner edge (centripetally) and secondary phloem at its outer edge (centrifugally), although plants with anomalous secondary growth do not always follow this pattern.

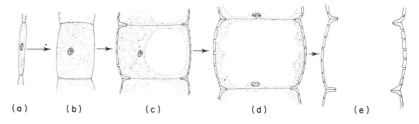

Fig. 2.7   Ontogeny of a vessel element **(e)** from a fusiform initial **(a)** in *Robinia pseudoacacia*. (After Eames and MacDaniels, 1925.)

The vascular cambium is a complex tissue consisting of both fusiform initials and ray initials, which form the axial and radial systems respectively. Fusiform initials are axially elongated cells with tapering ends (Fig 2.7). They divide periclinally to form the axial elements of secondary tissues: tracheary elements, fibres and axial parenchyma in secondary xylem, and sieve elements, companion cells and fibres in secondary phloem. Ray initials are more or less isodiametric cells that divide periclinally to form ray parenchyma cells in both xylem and phloem. Fusiform initials sometimes give rise to new ray initials as the stem increases in circumference, and new rays are formed.

Secondary xylem of woody dicotyledons (known commercially as hardwoods) varies considerably in texture, density and other properties, depending on the size, shape and arrangement of the constituent cells (Jane, 1970;

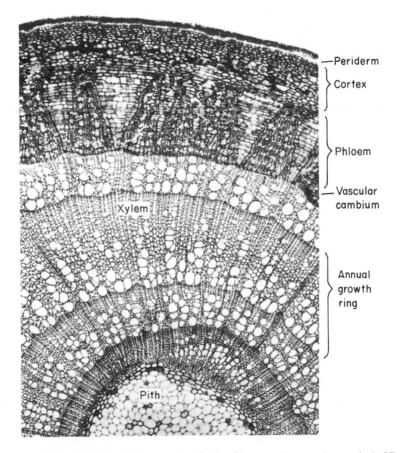

Fig. 2.8 *Tilia olivieri* (lime). Cross section of twig of just over three year's growth. (×25)

Metcalfe and Chalk, 1983). Wood is composed of a matrix of cells (Fig. 2.9), some arranged parallel to the long axis (fibres, vessels and chains of axial parenchyma cells), and others (ray parenchyma cells) forming the wood rays that extend radially from the vascular cambium towards the pith. The variation between different types of wood ensures that many woods can be fairly accurately identified by their anatomical structure (Miles, 1978; Schweingruber, 1978). To observe their structure, woods are cut in cross section and in two longitudinal planes of section: along the rays (radial longitudinal section) and at right angles to the rays (tangential longitudinal section). In Fig. 2.9 a cross section and a tangential longitudinal section can be seen. In oak the vessels are solitary in cross section, but in other woods they may be in clusters or radial chains. Axial parenchyma cells may be independent of the vessels (apotracheal) or associated with them (paratracheal), and are sometimes in regular tangential bands. Relative abundance of axial parenchyma varies, from sparse – or even completely absent – to rare cases such as balsa wood (*Ochroma pyramidale*), where axial parenchyma cells are often more abundant than fibres. Variation in ray structure is best observed in longitudinal (particularly tangential) sections. Rays are termed uniseriate if they are one cell wide tangentially (Fig. 2.9), and multiseriate if they are more than one cell wide. Sometimes both uniseriate and multiseriate rays occur in the same wood, for example in oak (*Quercus* spp.). Ray cells vary in shape; homocellular rays are composed of ray cells of similar shapes, whereas in heterocellular rays the cells are of different shapes.

Other aspects of variation in the structure of hardwoods include the occurrence of either axial or radial secretory canals, the storied (stratified) appearance of various elements, particularly rays, in some woods, and the fine structure of the vessel walls (intervascular pitting, perforation plates and wall thickenings; section 1.5). For example in some woods, such as cherry (*Prunus* spp.) or lime (*Tilia cordata*, Fig. 1.4), the vessel element walls are helically thickened, and in other woods (e.g. many Leguminosae) the pit apertures are surrounded by numerous warty protuberances, known as vesturing. Perforated ray cells, an unusual and interesting feature of some woods, are ray cells that link two vessel elements and themselves resemble and function as vessel elements, with perforation plates corresponding to those of the adjacent vessel elements. They occur sporadically in some woods, and serve as an example of the adaptable nature of the vascular cambium, since they are formed from ray initials.

Many woody temperate plants have seasonal (usually annual) cambial activity, which results in the formation of growth rings. The secondary xylem formed in the early part of the season (early wood or spring wood) is generally less dense and consists of thinner walled cells than the xylem formed later in the growing season (late wood or summer wood). In ring porous woods the vessels are considerably larger in early than in late wood, although in diffuse porous woods the main distinction is in size and wall thickness of the fibres. As woody plants age and their trunks increase in girth, the central area becomes non-functional with respect to water trans-

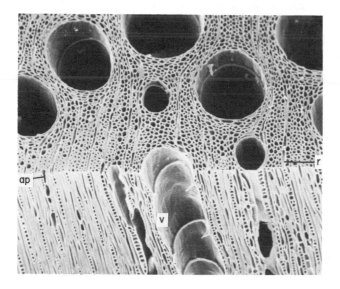

Fig. 2.9 *Quercus robur* (oak). Block of wood at edge of transverse and tangential longitudinal surfaces, showing large solitary vessels (v), uniseriate (rarely biseriate) rays (r), scattered axial parenchyma (ap), and fibres. (SEM, ×78)

port or food storage, and often the vessels become blocked by tyloses. Tyloses are formed when adjacent parenchyma cells grow into the vessels through common pit-fields. The central non-functional area of the trunk, the heartwood, is generally darker than the outer living sapwood.

There are many examples of woody dicotyledons, particularly climbing plants (lianes), in which secondary growth does not fit the 'normal' pattern of xylem and phloem production. These unusual forms, which occur in many plant families, are referred to as having anomalous secondary growth. For example, some plants have areas of phloem (included or interxylary phloem) embedded in the xylem, either in islands (e.g. in *Avicennia*), or in alternating concentric bands (Fig. 2.10d, f). Other examples may have irregularly divided or deeply fissured areas of xylem and phloem (Fig. 2.10b, e), a flattened stem (Fig. 2.10d), a deeply divided and irregularly shaped stem (Fig. 2.10c), or a quadrangular stem with the xylem deeply furrowed by regular areas of xylem (Fig. 2.10a) (Metcalfe and Chalk, 1983; Obaton, 1960; Schenck, 1893). These anomalous forms are achieved either by the formation of new vascular cambia in unusual positions, or by the unusual behaviour of the existing cambium, in producing phloem instead of xylem at certain positions.

Secondary phloem in dicotyledons, also a product of the vascular cambium, similarly displays great variation in structure. However this variation has been the subject of far fewer studies than secondary xylem, and

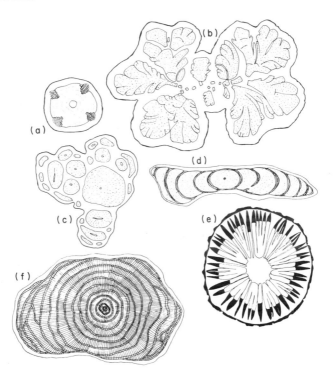

Fig. 2.10 Examples of anomalous secondary growth. (a) *Cuspidaria* sp. (Bignoniaceae), quadrangular stem with deeply furrowed xylem. (b) *Bauhinia* sp. (Leguminosae), irregularly shaped stem with fissured xylem. (c) *Thinouia* sp. (Sapindaceae), irregularly shaped stem with separate areas of vascular tissue. (d) *Machaerium* sp. (Leguminosae), flattened stem with alternating bands of xylem and phloem. (e) *Bidens* sp. (Compositae), deeply divided xylem. (f) *Abuta* sp. (Menispermaceae), concentrically alternating rings of xylem and phloem. (After Schenck, 1893.)

is consequently less readily used as a means of identification. As in secondary xylem, secondary phloem consists of both axial and radial systems, formed from the fusiform and ray initials respectively. Phloem rays are continuous with xylem rays, and may similarly be uniseriate or multiseriate, although they are often dilated towards the cortex as a result of cell divisions to accommodate increase in stem thickness (Fig. 2.8). The parenchymatous ray cells are often difficult to distinguish from cortical cells at their junction, and older ray cells sometimes become lignified to form sclereids. The axial system of the phloem consists of sieve elements and companion cells, as in primary phloem (section 1.5), and also often includes fibres, sclereids and axial parenchyma cells. In some cases, such as lime (*Tilia* sp., Fig. 2.8), the fibres are formed in groups at regular intervals, resulting in characteristic tangential bands of fibres alternating with areas of sieve elements and parenchyma cells.

2.3.2   Stem thickening in monocotyledons

In many herbaceous monocotyledons there is little or no stem thickening growth, but in cases where increase in thickness does occur, it is initiated at the primary thickening meristem near the vegetative shoot apex (Fig. 2.11). In very few monocotyledons further increase in thickness also occurs at a greater distance from the apex, either by diffuse secondary growth, or at the secondary thickening meristem (Fig. 2.12).

Most monocotyledons which have thick stems with short internodes and crowded leaves have primary thickening meristems. These are situated in the pericyclic region just below the apex, and consist of a relatively narrow zone of meristematic cells producing radial derivatives, usually parenchyma towards the outside (centrifugally), and both parenchyma and vascular tissue towards the inside (centripetally). Vascular bundles including both xylem and phloem are formed from centripetal derivatives of the meristem.

In many monocotyledons the primary thickening meristem ceases activity at a short distance behind the apex, and subsequent stem thickening is limited. In palms however, considerable stem thickening occurs by extensive division and enlargement of ground parenchyma cells. This is termed diffuse secondary growth. In some woody Liliflorae, such as species of *Aloë*, *Yucca* and *Agave*, further increase in stem thickness is achieved by secondary growth from a secondary thickening meristem.

The secondary thickening meristem is essentially similar to the primary thickening meristem in that it is situated in the pericyclic region and produces similar radial derivatives, but differs in that it occurs further from the stem apex, and the secondary vascular bundles produced are often amphivasal and radially elongated. The similarity between the two meristems, and

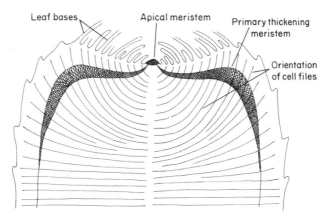

Fig. 2.11   Diagram of longitudinal section of the crown of a typical thick-stemmed monocotyledon with a primary thickening meristem, showing orientation and extent of radial derivatives. Vascular strands not shown. (Adapted from DeMason, 1983.)

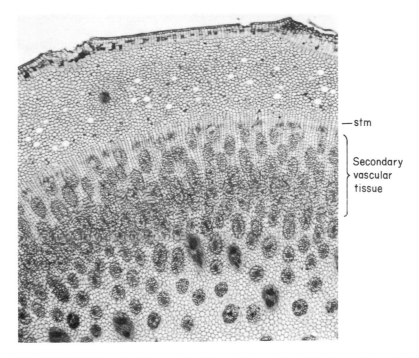

—stm

} Secondary
vascular
tissue

Fig. 2.12   *Cordyline rubra.* Cross section of stem with secondary thickening meristem (stm). (×20)

the fact that they are sometimes continuous (for example in *Yucca whipplei*) has led some authors (such as Diggle and DeMason, 1983) to regard them as developmental phases of the same meristem. Although Stevenson (1980) and Stevenson and Fisher (1980) showed that in mature stems of *Beaucarnea recurvata* and *Cordyline terminalis* the primary and secondary thickening meristems are often longitudinally discontinuous, Diggle and DeMason (1983) suggested that this discontinuity is caused because periclinal divisions in the meristem decrease during stem elongation and subsequently resume.

The primary and secondary thickening meristems of monocotyledons cannot be considered as homologous to the vascular cambium of dicotyledons because the vascular derivatives are arranged in completely different ways. In dicotyledons the vascular cambium produces phloem centrifugally and xylem centripetally, whereas in monocotyledons discrete vascular bundles of both xylem and phloem are formed centripetally. Also, as DeMason (1983) pointed out, near the shoot apex the primary thickening meristem represents a relatively diffuse zone of meristematic activity compared to the uniseriate vascular cambium of dicotyledons.

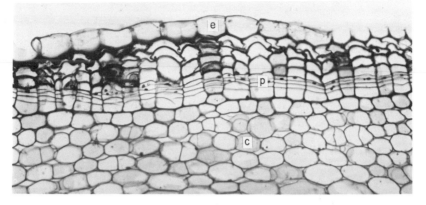

Fig. 2.13 *Sambucus nigra* (elder). Cross section of stem surface, showing broken epidermis (e), with periderm (p) forming beneath, in outer cortical layers: c = cortex. (×130)

2.4 Periderm

Periderm is a protective tissue of corky (suberized) cells that is often produced as a response to wounding. It consists of up to three layers: phellogen, phellem and phelloderm. The phellogen is a uniseriate meristematic layer of thin walled cells which produces the other two layers, the phellem to the outside, and (in some cases) the phelloderm to the inside. The phellem cells constitute the corky tissue. They are tightly packed cells devoid of contents at maturity. They have deposits of suberin and sometimes lignin in their walls, and form an impervious layer to prevent water loss and protect against injury. Phelloderm cells are non-suberized and parenchymatous, and contribute to the secondary cortex.

A periderm commonly occurs in the cortex of secondarily thickened stems, to replace the epidermis, which splits and peels away. The phellogen may initially form either adjacent to the epidermis (or even in the epidermis) or deeper in the cortex. Sometimes several phellogens may form almost simultaneously. The pattern of periderm formation largely dictates the appearance of the bark of a woody plant (the term 'bark' usually denoting all the tissues outside the vascular cambium). For example, the smooth papery bark of a young birch tree is formed because the periderm initially expands tangentially with the increase in stem diameter, but later flakes off in thin papery sheets as abscission bands of thin walled cells are formed. In the trunk of the cork oak (*Quercus suber*), the source of commercial cork, the initial phellogen may continue activity indefinitely and produces seasonal growth rings, although in the commercial process it is removed after about 20 years to make way for a second, more vigorous phellogen, which provides the commercial cork. Many barks have lenticels, which are areas of loose cells in the periderm, often initially formed beneath stomata in the epidermis, and sometimes thought to be similarly concerned with gaseous exchange.

3

The Root

3.1 Root apex

Root apices have a terminal protective root cap composed of several layers of parenchymatous cells. The root apical meristem is adjacent and proximal to the root cap, and the junction between the cap region and the meristem may either be clearly defined by a distinct cell boundary ('closed' structure, as in *Zea mays*, Fig. 3.1), or ill-defined ('open' structure, as in *Vicia faba*, Fig. 3.2). Views on the organization of the root apical meristem have in the past differed widely (Clowes, 1961). However the discovery of the 'quiescent centre', a group of relatively inactive cells at the very centre and tip of the apical meristem, has led to an adjustment in our understanding of the zonation of mitotic activity. Contemporary opinions suggest that most cell division activity occurs in the area of cells proximal to the quiescent centre, which produce the bulk of the root tissue. This active region is termed the promeristem. The cells of the quiescent centre divide infrequently, and its role is somewhat obscure, although many investigators believe that it influences the cellular organization of the root (Feldman, 1984). The quiescent centre is defined only by cell division activity, and is not always distinct in root sections.

The cells of the root cap are also initially derived from the apical meristem. However ontogenetic studies on *Zea mays*, a species with 'closed' root apical structure, have shown that the cap initials become established and independent from the apical meristem at an early stage in the development of the seedling (Barlow, 1975). The cap meristematic cells, located adjacent (distal) to the quiescent centre, produce derivatives which are eventually displaced towards the outside of the root cap, and subsequently sloughed off, contributing to the external 'slime'. Cells are generated and lost in the root cap at approximately the same rate.

3.2 Primary root structure

Although roots can originate from various organs in the plant (section 3.3), basic primary structure is relatively uniform, and different to that of

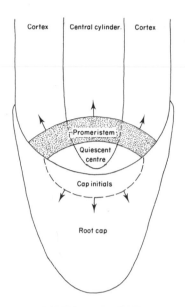

Fig. 3.1 Diagram of organization of root apex in *Zea mays* (maize), a species with 'closed' structure. Arrows indicate direction of displacement of cell derivatives. (Adapted from Feldman, 1984.)

the stem. Each root has an epidermis, cortex and central vascular region. The central region is usually bounded by an endodermis and pericycle (Fig. 3.3).

3.2.1 Root epidermis
The root epidermis is typically uniseriate, as in other parts of the plant. In underground roots it often becomes worn away, and is replaced as an outer protective layer either by a periderm (in most woody dicotyledons), or an exodermis. In a few monocotyledons, particularly those with aerial roots, such as some tropical Orchidaceae, the root epidermis is multiseriate, termed a velamen, which is commonly believed to have a function in water absorption, by retaining water after rain.

Most angiosperms have root hairs, formed by extension of epidermal cells. In some plants only certain epidermal cells, the trichoblasts, are capable of root hair production. Root hairs usually arise about a centimetre from the root apex, beyond the meristematic region, but in an area where cells are still

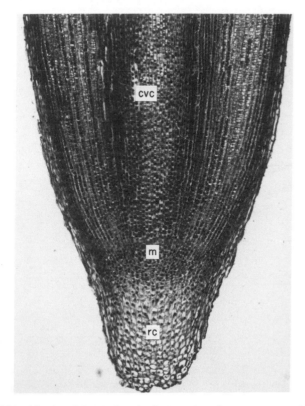

Fig. 3.2   *Vicia faba* (broad bean). Longitudinal section of root apex; 'open' structure: cvc = central vascular cylinder, m = meristem, rc = root cap. (×60)

enlarging. In general they persist for only a limited period before withering. This region of the root is the most active in absorption of water, and the root hairs serve to present a greater surface area for this purpose.

3.2.2   Cortex

The cortex constitutes the region between the pericycle and the epidermis, including the exodermis, where present, and also the endodermis, which is derived from cortical cells. Apart from these specialized layers most cortical cells are parenchymatous and often perform an important storage function. In some plants, such as *Daucus carota* (carrot), the tap root is a modified swollen storage organ with a wide cortex. In most roots the bulk of cortical cells are formed by sequential periclinal divisions, the innermost cells (later the endodermis) being the last formed.

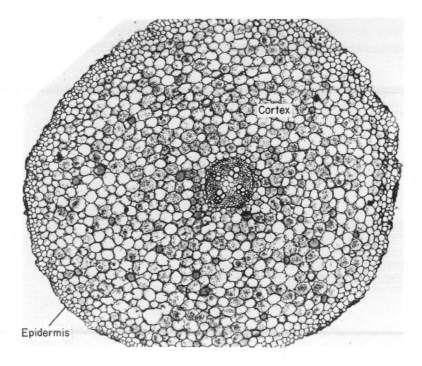

Fig. 3.3 *Ranunculus* sp. Cross section of root. (×60)

Many plants with underground stems (corms, bulbs or rhizomes), particularly bulbous or cormous monocotyledons such as *Crocus, Freesia* or *Hyacinthus*, periodically produce contractile roots which draw the stem deeper into the soil (Fig. 3.4). Root contraction is brought about by radial expansion and usually longitudinal shortening of inner or middle cortical cells, often accompanied by the passive collapse of outer cortical parenchyma (Jernstedt, 1984).

The exodermis, which occurs in roots of a few monocotyledons, such as *Iris*, is formed from subepidermal cortical cells, usually one to several layers, which become suberinized or lignified as the epidermis is worn away in older roots.

The endodermis is a uniseriate cylinder of cortical cells surrounding the central vascular tissue, adjacent to the pericycle. These cells have become modified, typically by deposition of a band of suberin or lignin (Casparian strip) in their primary walls, forming a barrier against non-selective passage of water through the endodermis. Older endodermal cells often have thick lamellated secondary walls, in most cases on the inner periclinal walls, so

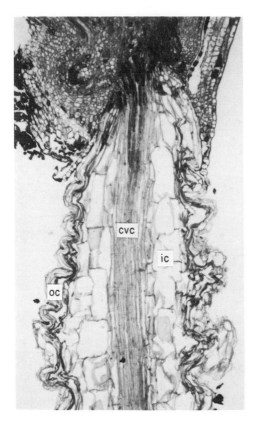

Fig. 3.4 *Tigridia dugesii.* Longitudinal section of contractile root at base of bulb, showing central vascular cylinder (cvc), expanded inner cortical cells (ic) and collapsed outer cortical cells (oc). (×45)

that the Casparian strip is not apparent. The secondary wall deposits are often lignified, and also serve as an effective barrier to water. Occasional endodermal cells (passage cells) may remain thin walled, probably for passage of material between the cortex and vascular region.

3.2.3 Central vascular cylinder

The vascular tissue in the centre of the root is surrounded by one or more layers of thin walled cells which constitute the pericycle. The pericycle is potentially meristematic in younger roots, as it is the site of lateral root initiation, but in older roots it is less active and may become lignified. The primary vascular tissue consists of separate strands of phloem alternating with the 'arms' (archs) of a central area of xylem which appears star-shaped in cross section. The protoxylem elements, which are the first-formed and narrowest in diameter, are at the tips of the 'arms', nearest to the pericycle,

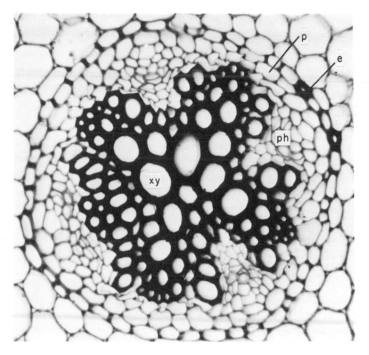

Fig. 3.5 *Ranunculus acris* (buttercup). Cross section of central part of root without secondary thickening. Xylem (xy) with four protoxylem poles; four areas of phloem (ph); pericycle (p); endodermis (e). (×380)

with the wider metaxylem elements closer to the centre of the root. Similarly, as both xylem and phloem are exarch in the root (i.e., mature centripetally), the protophloem is closer than the metaphloem to the pericycle. In a few monocotyledons, such as *Iris*, the xylem does not extend to the centre of the root, and the central tissue is parenchymatous, sometimes becoming lignified in older roots. More commonly the central region is occupied by a group or ring of large metaxylem vessels. Roots may have two, three or more xylem poles, in which cases they are diarch, triarch or polyarch respectively. However there is often variation in the number of xylem poles, even sometimes in the same plant, depending on the diameter of the root. Dicotyledonous roots usually have a relatively low number of xylem poles, most commonly two, three or four, but many monocotyledons have a much higher number.

3.3 Initiation of lateral and adventitious roots

Lateral roots are branches of the tap root, and have a deep-seated (endogenous) origin. Although the most recently formed lateral roots are usually

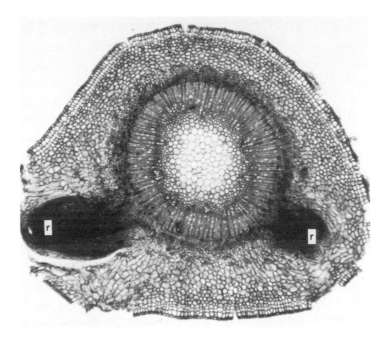

Fig. 3.6 *Ligustrum vulgare* (privet). Cross section of stem with adventitious roots (r). (×40)

those nearest to the root apical meristem, they are initiated in mature tissues some distance from the apex, often in acropetal sequence. In angiosperms the initial cell divisions in lateral root formation usually occur in the pericycle. This forms a lateral root primordium, with in many cases some subsequent cell divisions in the endodermis, so that ultimately both the pericycle and the endodermis contribute to the tissues of the lateral root. The growing lateral root pushes its way through the cortex and epidermis of the parent root, either by mechanical or enzymatic action. In dicotyledons the position of lateral root initiation in the pericycle is usually at a point adjacent to a protoxylem pole, unless the root is diarch, in which case initiation is sometimes opposite a phloem pole. In monocotyledons lateral root initiation can be opposite either protoxylem or phloem poles, although in roots with a large number of vascular poles it is often difficult to determine the precise site of initiation (McCully, 1975).

The term adventitious root, used in its broadest sense, refers to any root not arising from the seedling radicle or its branches (Esau, 1965). Adventitious roots may be deep-seated (endogenous) in origin (Fig. 3.6), or more rarely exogenous, arising from superficial tissues such as the epidermis. In most monocotyledons adventitious roots arise from cell divisions in the pericyclic region of the stem, and sometimes the primary thickening meristem may contribute to adventitious root formation. Adventitious roots are often

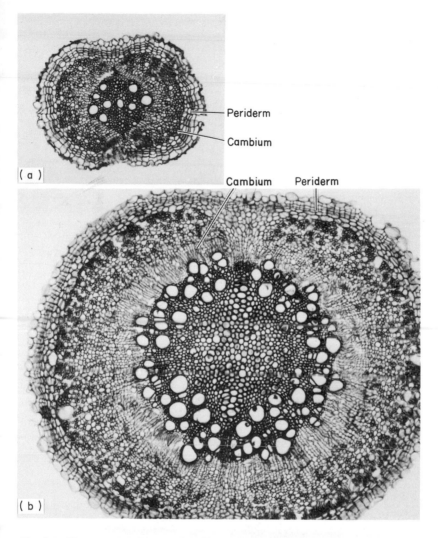

Fig. 3.7 *Ulmus* sp. (elm). Cross section of roots with (a) secondary growth just begun, and (b) secondary thickening more extensive. (×60)

formed at nodes on the stem, which is why cuttings are commonly taken from just below a node.

3.4 Secondary growth in roots

In monocotyledons secondary growth in roots is extremely rare, even among species which have secondary growth in the stem (section 2.3.2). Among arborescent monocotyledons it is confined to roots of *Dracaena* (Tomlinson and Zimmermann, 1969), and even here it is somewhat limited. It initiates in the pericycle or cortex, or sometimes both, and the secondary vascular tissue is similar to that of the stem.

Most dicotyledonous roots have a certain amount of secondary thickening, with the exception of a few herbaceous species, such as *Ranunculus* (Figs. 3.3, 3.5). In some tree species the thickening and strengthening of the root system may be important in supporting the trunk. As in the stem, the secondary vascular tissues of the root are produced by a vascular cambium. This initially develops in the regions between the primary xylem and phloem, then in derivatives of cell divisions in the pericycle next to the xylem poles. Since cambial activity proceeds in this sequence, the xylem cylinder soon appears circular in cross section (Fig. 3.7). Further pericyclic cell divisions result in the formation of a secondary cortex, and in many cases, particularly where secondary growth is extensive, the primary cortex and endodermis become split and are sloughed off, and a periderm forms.

Root secondary xylem is usually fairly similar to that of the stem in the same plant, but many differ in several respects. For example in ring porous species of *Quercus* (oak), such as *Q. robur*, the root wood is diffuse porous, the

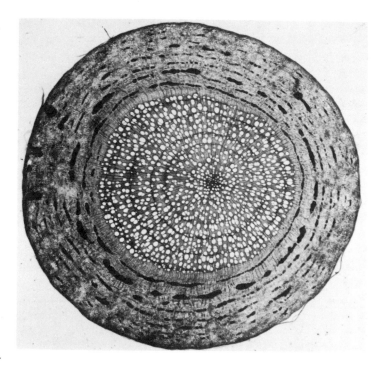

Fig. 3.8 *Populus tremula* (aspen). Cross section of woody root. Secondary phloem with concentrically arranged blocks of fibres; secondary xylem with four growth rings. (×10)

vessels being similar in size across each growth ring. This is in contrast to the stem, where the early wood vessels are markedly larger than the late wood vessels.

3.5 Haustoria of parasitic angiosperms

There are many examples of angiosperms that are parasitic on the roots, stems or leaves of other angiosperms. Perhaps the best known are the mistletoes (e.g. *Viscum album*, family Loranthaceae), but others include the dodder (*Cuscuta* spp., family Convolvulaceae), sandalwoods (family Santalaceae), broomrapes and parasitic figworts (families Orobanchaceae and Scrophulariaceae). The haustorium is a highly modified root through which nutrients flow from the host to the parasite. Haustoria may be termed primary or secondary; a primary haustorium is a direct outgrowth of the apex of the parasite radicle, whereas a secondary haustorium is a lateral organ which may form from a modified adventitious root or from outgrowths of roots or stems (Kuijt, 1969). Haustoria take many different forms

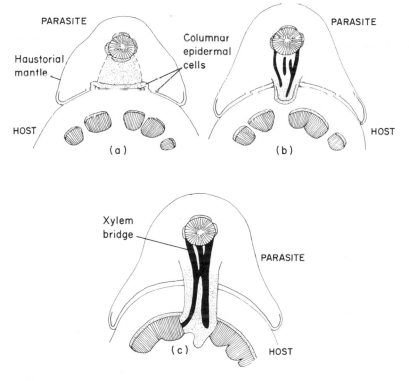

Fig. 3.9 Diagram to show penetration of host tissue by parasitic haustorium of *Cassythia* sp. (Adapted from Kuijt, 1969.)

in different parasites. They often penetrate the host tissue to the xylem and form a continuous xylem bridge between the host and parasite. The phloem tissues are not linked in the same way, except sometimes by specialized parenchyma cells. In sandalwoods and many other parasites the developing haustorium forms a 'mantle' of mainly parenchymatous tissue around the host organ. The epidermal cells in contact with the host become elongated and secretory, and the centre of the haustorium develops an intrusive process which grows into the host by enzymatic and mechanical action (Fig. 3.9). In Loranthaceae the haustorium often does not invade the host tissues after forming a mantle, but instead influences the host tissue to form a placenta-like outgrowth of vascular tissue to supply nutrients to the parasite. When the parasite dies a woody outgrowth of convoluted host tissue remains, known as a 'woodrose'.

4

The leaf

The mature lamina consists of an adaxial and abaxial epidermis enclosing several layers of mesophyll cells interspersed with a network of vascular bundles. Each tissue may be variously differentiated into specialized cell types, the degree of differentiation varying considerably between taxa. Angiosperm leaves display much diversity in cross-sectional outline. In dorsiventral (bifacial) leaves the upper and lower surfaces differ, for example in relative numbers of stomata and trichomes, and the mesophyll may be differentiated into both palisade and spongy tissues (Fig. 4.1). In isobilateral (unifacial) leaves (Fig. 4.2) the epidermis and mesophyll are usually more or less similar on both surfaces, and in centric leaves (Fig. 4.8) they are continuous. Some monocotyledons, such as *Libertia* (Fig. 4.2) have unifacial leaves with a bifacial sheathing leaf base.

4.1   Leaf development

Leaves are initiated as small conical projections (leaf primordia) close to the stem apex, by periclinal cell divisions either in the outermost cell layers, or in the layers immediately below them. Elongation continues by apical

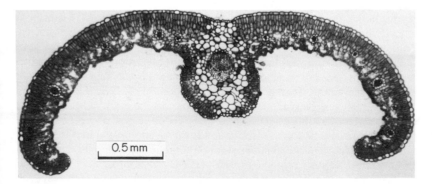

Fig. 4.1 *Helichrysum italicum.* Leaf, cross section.

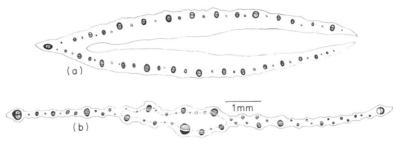

Fig. 4.2 *Libertia chilensis.* Cross section of leaf. **(a)** Bifacial leaf base. **(b)** Isobilateral leaf blade.

growth for a short time in the primordia, then it is replaced by intercalary growth. In young primordia the adaxial marginal cells divide rapidly to form a flattened leaf blade. This marginal growth is suppressed in the area which later becomes the petiole, and in many monocotyledons it is often simultaneous with apical growth. Marginal growth is usually subsequently replaced by cell divisions across the whole leaf blade. By this stage the number of cell layers comprising the leaf thickness has been more or less established, and the whole lamina functions as a plate meristem. Cell divisions are mainly anticlinal, resulting in regular layers of cells which are disrupted only by the differentiation and maturation of the vascular bundles. Rates of growth and cell division may sometimes vary in different parts of the leaf. In many monocotyledons meristematic activity governing leaf elongation is restricted to a meristem at the base of the leaf, the basal rib meristem, which results in longitudinal files of cells of increasing maturity towards the distal end of the leaf (see also section 4.2.2.).

Several investigators (e.g. Kaplan, 1970, 1973a,b, 1975, 1984) have attempted to relate differences in histogenesis in leaves to differences in mature leaf structure in various plant taxa. For example, the unifacial leaves of some monocotyledons with a bifacial sheathing base and a unifacial upper blade, have long been interpreted as resulting from suppressed marginal growth and emphasized adaxial growth. By means of comparative studies, both morphological and anatomical, Arber (1925) and others have contended that the unifacial leaves of some monocotyledons are homologous with the phyllodes of some dicotyledons such as *Acacia*, and that both are derived from a petiolar structure, after suppression of the leaf blade. More recently however, Kaplan (1975), by comparing the early development of several unifacial monocotyledon and dicotyledon leaves, accepted that the two structures are homologous, but as a result of an unusual method of leaf expansion by a single adaxial meristem rather than from petiolar derivation. In another investigation Kaplan (1984) examined developmental mechanisms in other unrelated plants with similar structural features, such as dissected or lobed leaves in the families Araceae and Palmae, and found very different ontogenetic sequences. As a result of these and other investigations the advantages of an ontogenetic approach to evolutionary problems are now widely recognized.

4.2 Leaf epidermis

The epidermis usually consists of a single layer of cells, although in rare cases (such as species of *Ficus* and *Peperomia*) it may have proliferated to form a multiple epidermis. This is similar in structure to a hypodermis (section 4.3), and can be distinguished from it only by a developmental study.

The specialized elements of the leaf epidermis are essentially the same as those of the stem: stomata, trichomes, papillae, surface sculpturing, epicuticular wax and variously differentiated epidermal cells. However the leaf surface has been the subject of more investigations than other plant surfaces, and since many of the variable features are constant within taxa they often have taxonomic applications.

4.2.1 Epidermal cells
In surface view epidermal cells may be elongated or more or less isodiametric, and the anticlinal walls (at right angles to the plant surface) may be straight (Fig. 4.3b) or undulating (Fig. 4.3a, c). Cells over veins are often elongated in the direction of the vein, and in long narrow leaves of the sort found in many monocotyledons, such as *Iris*, epidermal cells are frequently elongated along the long axis of the leaf. The anticlinal walls are often more sinuous on the abaxial than the adaxial surface of the same leaf.

Epidermal cells may also vary in size and wall thickness in different parts of the same leaf. A particularly striking example of this occurs in some Gramineae, such as *Zea mays*, where bulliform cells occur in restricted areas of the leaf epidermis. These cells are larger than other epidermal cells and thin walled, and may in some cases have a function in the unrolling of the leaf in response to turgor pressure and water availability.

Occasionally epidermal cells may contain crystals or silica bodies. Characteristic silica bodies are found in the leaf epidermis of many species, particularly in the monocotyledon families Cyperaceae, Gramineae and Palmae. In the Gramineae the epidermis typically consists of both long and short cells, the short cells sometimes containing silica bodies or forming the bases of hairs.

4.2.2 Stomata
Stomata are the pores in the epidermis through which gaseous exchange takes place. They occur on most of the plant surfaces above ground, especially on green photosynthetic stems and leaves, but also on floral parts. On leaves stomata may be found on both surfaces (amphistomatic), or on the abaxial surface only (hypostomatic), or in rare cases, such as on floating leaves, on the adaxial surface only (epistomatic). They occur most

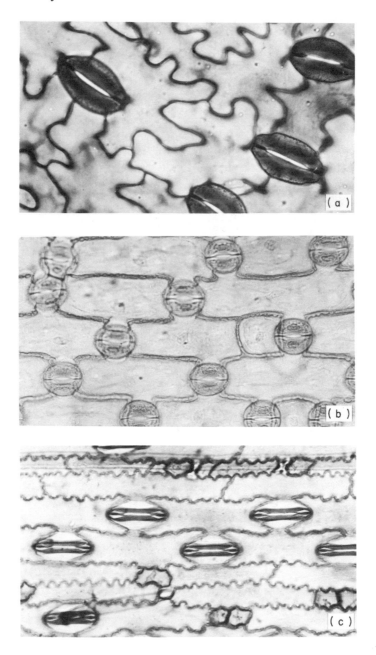

Fig. 4.3 Abaxial leaf surfaces with stomata. **(a)** *Paeonia officinalis*, **(b)** *Crocus speciosus*, **(c)** *Arundo donax*. (×380)

frequently on the regions overlying the mesophyll rather than over the veins, and may sometimes be restricted to certain areas, such as in grooves or depressions.

Each stoma consists of two kidney shaped guard cells surrounding a central pore. Cuticular ridges may extend over or under the pore from the outer or inner edges of the adjacent guard cell walls. Stomata may be sunken or raised, and are usually associated with a substomatal cavity in the mesophyll.

The epidermal cells immediately adjacent to the guard cells are termed subsidiary cells if they are different in shape to the surrounding epidermal cells. Classifications of stomatal types are based either on the arrangement of mature subsidiary cells, or on their patterns of development. The most widely used system of classification of mature stomatal types includes many terms such as: anomocytic, with no subsidiary cells; anisocytic, with three subsidiary cells; diacytic, with one or more pairs of subsidiary cells with their common walls at right angles to the guard cells; and paracytic, in which, at either side of the guard cells, there are one or more subsidiary cells which may or may not meet at the poles. (See Wilkinson, 1979, for a summary of mature stomatal types). However, as Tomlinson (1974) and others have pointed out, this classification, although useful in a descriptive sense, does not take into account the different developmental pathways which may lead to similar stomatal types, and therefore has a limited application.

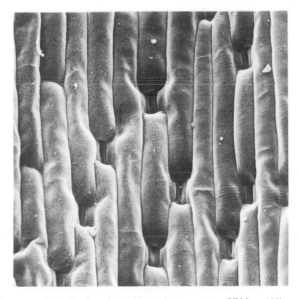

Fig. 4.4 *Crocus speciosus*. Leaf surface with sunken stomata. (SEM, ×400)

The importance of ontogenetic studies in comparing stomata of different taxa is now widely recognized. Stomatal ontogeny is particularly accessible in monocotyledons as their leaves grow with a basal rib meristem which often produces longitudinal files of epidermal cells with a sequence of development along the leaf (Tomlinson, 1974; Rasmussen, 1983). In monocotyledons there is an unequal mitotic cell division which produces a larger cell and a smaller guard-cell-mother-cell (meristemoid). If the development is agenous the meristemoids give rise directly to the guard cells and there are no subsidiary cells. In perigenous modes of development the meristemoid gives rise directly to the guard cells and the subsidiary cells are formed from neighbouring cells, often by oblique divisions. In mesogenous modes of development the meristemoid undergoes a further mitotic division into two daughter cells before one cell further subdivides to form the guard cells, the other cell usually forming one or more subsidiary cells. Subsidiary cells derived from the meristemoid are termed mesogene cells, whereas those derived from the neighbouring cells are termed perigene cells. In some cases mesogene cells are not distinct from surrounding epidermal cells at maturity.

4.2.3 Trichomes and papillae

Trichomes are epidermal outgrowths which occur on all parts of the plant surface (Fig. 4.5). They vary widely in both form and function, and include unicellular or multicellular, branched or unbranched forms, and also scales, glandular (secretory) hairs, hooked hairs and stinging hairs. The distinction between trichomes (hairs) and papillae, which are also epidermal outgrowths, is not always clear, although papillae are generally smaller and unicellular. In cases where there are several small outgrowths on each epidermal cell these outgrowths are usually termed papillae, but where there is only one unicellular outgrowth per cell, the distinction is dependent on size.

Trichomes may occur on the entire leaf surface, or may be restricted to certain areas, such as in grooves (Fig. 4.12) or at the margins. Often there are several different types on the same leaf. For example, in many genera of Labiatae, such as *Rosmarinus* (rosemary), *Thymus* (thyme) and *Lavandula* (lavender), there are two or more sizes of glandular hair and branched or unbranched non-glandular trichomes.

Glandular trichomes usually have a one or more celled stalk and a secretory head with one to several cells. In some glandular hairs the substances secreted collect between the secretory cells and a raised cuticle, which may later break to release them. There are many different types of glandular hair, and they secrete a variety of substances, from salt or essential oils to digestive juices in some carnivorous plants (Fahn, 1979). The stinging hairs of *Urtica dioica* (stinging nettle) are rigid, hollow structures which contain a poisonous substance. The spherical tip of the hair is readily broken off in contact with an outside body, and the remaining sharp point may then penetrate the skin and release the fluid. Another example of a specialized hair type is the

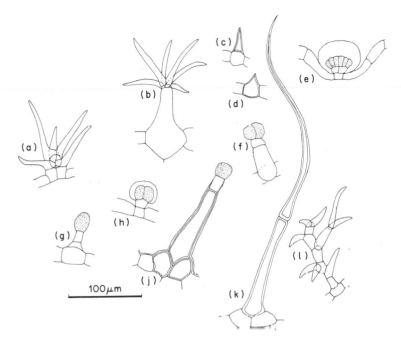

Fig. 4.5 Range of hair types in Labiatae. **(a)** *Phlomis fruticosa*, **(b)** *Lavandula* sp. **(c,d)** *Thymus vulgare*, **(e)** *Origanum* sp., **(f)** *Stachys sylvatica*, **(g,h,j)** *Coleus cosmosus*, **(k)** *Lamium album*, **(l)** *Rosmarinus officinalis*. **(e-i)** glandular hairs; **a-d, k, l** non-glandular hairs.)

scales on the leaves of many Bromeliads, which are specially adapted for water absorption.

4.2.4 Cuticle and wax

The cuticle, which is composed mainly of cutin, lipids and waxes, is a layer covering the entire leaf surface (and most other aerial plant surfaces). In mesomorphic leaves the cuticle is often thin and almost transparent; however many xeromorphic plants have thick cuticles which often appear lamellated in transverse section. In some plants the outer surface has characteristic patterns of cuticular ridges or folds. These striations may be short or long and with regular or random orientation, sometimes radiating from around stomata or trichomes. Barthlott and Ehler (1977) examined epidermal features of many angiosperms using a scanning electron microscope, and found much variation in both cell shape and surface sculpturing, often with taxonomic applications. They discussed the biological significance of surface sculpturing in relation to mechanical and optical properties and wettability of the surface. Cutler and Brandham (1977) and Brandham and Cutler (1978) showed that the cuticular patterns on the mature leaf surfaces of *Aloë* and related genera are uniform within species and under precise genetic control.

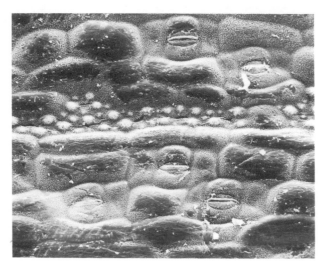

Fig. 4.6 *Gibasis pulchella*. Leaf surface, showing stomata, epicuticular wax particles and surface markings. (SEM, ×300)

Leaves may also have a covering of wax over the cuticle. The wax is either in the form of a surface crust or more commonly in small particles of varying shapes and sizes, ranging from flakes to filaments and granules. Wax particles are variously orientated and may be in characteristic patterns. Barthlott and Wollenweber (1981) showed that many aspects of the morphology and chemistry of epicuticular waxes have taxonomic applications, particularly the orientation and distribution patterns of the particles. To some extent certain compounds, such as terpenes and flavonoids, can be recognized by their fine structure using the scanning electron microscope (see also Juniper and Jeffree, 1983).

4.3   Mesophyll

Chlorophyll is contained in chloroplasts in the mesophyll, which is the primary photosynthetic tissue of the leaf. The mesophyll is often divided into palisade and spongy tissues, although it may sometimes be relatively undifferentiated and homogeneous throughout the leaf. Cells of palisade mesophyll, which is usually adaxial, are anticlinally elongated and with few intercellular air spaces, whereas spongy mesophyll, which is most commonly abaxial, consists of variously shaped cells with many air spaces between them. Both palisade and spongy tissues may be from one to several cell layers thick, and there is sometimes intergradation between the two tissues. Many tropical grasses, and also other unrelated taxa, have a ring of mesophyll cells radiating from the vascular bundles. This structure is com-

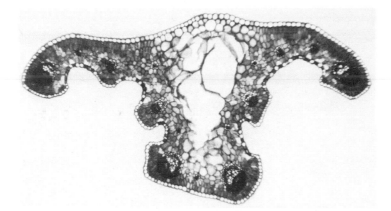

Fig. 4.7 *Crocus cancellatus.* Cross section of leaf. (×40)

monly associated with the C4 pathway of photosynthesis (section 4.4). In thick leaves, particularly those of some monocotyledons, the central cells are often large, undifferentiated and non-photosynthetic. In the thick 'keel' or 'midrib' of *Crocus* leaves there is a region of large colourless cells with their walls often broken down to form a cavity (Fig. 4.7). This causes the characteristic white stripe in the 'midrib' region of *Crocus* leaves.

Some xeromorphic plants, such as species of *Ilex* and many other genera, have a hypodermis, which occurs immediately within the adaxial (or more rarely the abaxial) epidermis. This consists of one or more layers of non-photosynthetic cells which are usually slightly larger and thicker walled than the adjacent mesophyll cells, and often similar in transverse section to the epidermal cells, although without the specialized elements of the epidermis.

4.3.1 Sclerenchyma and idioblasts

Mesophyll may be interspersed with areas of sclerenchyma, particularly at the margins and extending as girders from the vascular bundles to the epidermis. Sclerenchyma, usually a strengthening tissue, consists of thick walled lignified cells which generally fall into one of two categories: fibres, which are elongated and generally pointed at both ends, or sclereids, which are isodiametric or variously shaped (section 1.5). Fibres are typically found in groups associated with the vascular bundles or margins, but sclereids are often isolated in the mesophyll, and may be classified into different types depending on their shape. For example, sclereids which are more or less star shaped with many arms (e.g. in leaves of *Nymphaea*) are astrosclereids, and those which are bone shaped with a central column and branched ends (e.g. in leaves of *Hakea*, Fig. 4.8) are termed osteosclereids. However many sclereids do not fit readily into any categories due to their often bizarre shapes. In some cases (e.g. in leaves of *Memecylon* and *Mouriri*) sclereids are

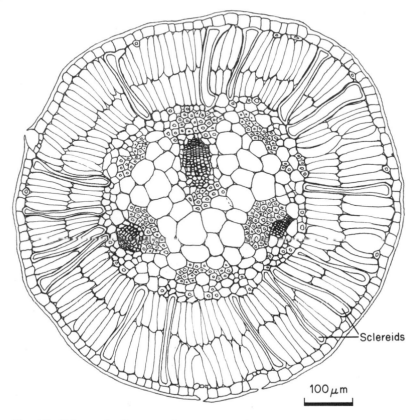

Fig. 4.8 *Hakea* sp. Leaf, cross section.

associated with veinlet endings. Many investigations have been carried out on sclereids in leaves of vascular plants, often with taxonomic applications (e.g. Foster, 1947, 1956. See Metcalfe and Chalk, 1979, for a survey of records of sclereids in dicotyledons). Highly branched sclereids grow intrusively into the intercellular air spaces between neighbouring cells in the developing leaf.

Other types of idioblasts (single modified cells) may also be interspersed in the mesophyll. For example, secretory myrosin cells are often found in the leaves of Cruciferae, and laticifers, which are individual cells or networks of cells secreting latex, occur in leaves of *Euphorbia*, *Allium*, and many other genera (Fahn, 1979). Other plants may have individual specialized mesophyll cells containing substances such as tannin, oil, mucilage or crystals.

4.4   Vasculature

The term venation is usually applied to the arrangement of the vascular bundles (veins). There are two main venation types among the angiosperms: parallel and reticulate, the former being broadly typical of monocotyledons and the latter of dicotyledons, although there are many exceptions to this.

In leaves with parallel venation the main veins are parallel for most of their length and merge or fuse at the leaf tip. There are numerous small veins interconnecting the larger veins but very few vein endings in the mesophyll. In leaves with reticulate venation (Fig. 4.9) there is often a major vein in the middle of the leaf, the midrib or primary vein, which is continuous with the major venation of the petiole, and has slightly smaller (second order) veins branching from it and often extending to the margins. Smaller veins may in their turn branch from the second and subsequent order venation, forming a network. The areas of mesophyll between the smallest veins in the leaf are

Fig. 4.9 *Hyptis vauthieri*. Cleared leaf with reticulate venation. (×5.6)

termed areoles, and in many cases small veins branch into the areoles to form vein endings. There are many variable aspects of venation: the relative number of veinlet endings per areole, or whether second order veins terminate at the margins or loop around to fuse with the superadjacent secondary veins. The patterns of venation in dicotyledons have been variously classified (e.g. by Hickey, 1973, 1979) and are often characteristic of species, genera or families.

Among reticulate veined dicotyledons the simplest form of petiole vasculature appears in cross section as a crescent, with xylem on the adaxial side and phloem on the abaxial side (Fig. 4.10). In some cases there may be additional bundles outside the main crescent, which may itself be inrolled at the ends, or in a ring, or divided into separate bundles. The classification of these various forms depends on how the petiole vasculature is linked to the vasculature of the stem at the node (Howard, 1974, 1979). One or more traces may depart from each gap in the stem vascular cylinder (section 2.2.1). The number and pattern of vascular bundles sometimes vary along the length of the petiole, and midrib vasculature, which is continuous with that of the petiole, is subject to similar variation.

In cross sections of the lamina vascular bundles are usually in a single row, although very thick leaves, such as in some species of *Agave*, may have two

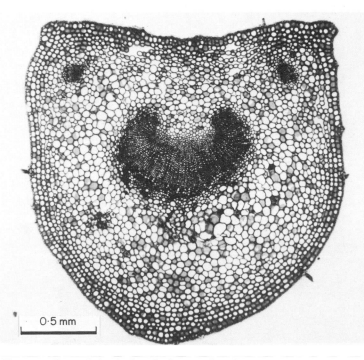

0·5 mm

Fig. 4.10 *Prunus lusitanica* (laurel). Cross section of petiole.

or more rows. Lamina bundles are usually collateral, with adaxial xylem and abaxial phloem, but orientation may vary, and in some cases bundles may be bicollateral or even amphivasal. In the isobilateral leaves of some monocotyledons (e.g. most Iridaceae, Fig. 4.2) there are two rows of vascular bundles with opposite orientation, the xylem poles towards the leaf centre. Centric leaves have a ring of vascular bundles (Fig. 4.8).

The xylem conducting system of the leaf blade may consist entirely of tracheids, usually with helical or annular thickenings, although in some leaves vessel elements and xylem parenchyma are also present. The smallest vascular bundles often have only one or two rows of xylem tracheids and a few files of phloem sieve tube elements.

Most minor vascular bundles in angiosperm leaves are surrounded by a bundle sheath, even to the very smallest veins. The bundle sheath typically consists of thin walled parenchymatous cells, often in a single layer and sometimes containing starch or chloroplasts. In some monocotyledons there are distinct inner and outer sheaths, the outer bundle sheath being parenchymatous and the inner sheath sclerenchymatous and often discontinuous, forming a sclerenchyma cap usually at the phloem pole. Grasses either have a single sheath of thin walled cells containing chloroplasts, or a double sheath consisting of an outer layer of thin walled cells and an inner layer of thicker walled cells. This is an important taxonomic character in the Gramineae, as double sheaths often occur in Festucoid grasses and single sheaths in Panicoid grasses, although there are occasional exceptions. Grasses from warm temperate areas with the C4 pathway of photosynthesis often display the 'Kranz' type of leaf anatomy, with mesophyll cells radiating from the vascular bundle sheaths, which consist of a single layer of thin walled cells

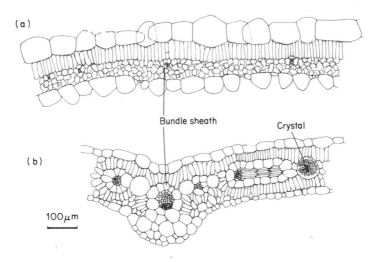

Fig. 4.11 Cross sections of leaves in region of secondary veins. (a) *Oxalis tuberosus.* (b) *Amaranthus caudatus*, leaf with Kranz anatomy.

containing chloroplasts and starch. Kranz leaf anatomy also occurs in many other plants with the C4 pathway of photosynthesis, including both dicotyledons and monocotyledons (Fig. 4.11b).

Leaves of many plants have areas of sclerenchyma or parenchyma extending from the vascular bundle sheaths towards either or both epidermises. These bundle sheath extensions, also called 'girders' if they reach the epidermis, afford mechanical support to the leaf and are often regarded as a xeromorphic feature (section 4.5).

4.5 Xeromorphy

Many anatomical features are termed xeromorphic because they are typical of plants which grow in dry (xeric) environments. Xeromorphic adaptations are generally considered to be effective in reducing water loss, protecting against excessive light intensity, and in affording mechanical support. Xeromorphic leaves are often very thick and sometimes also with reduced surface/volume ratio, although some xeromorphic plants have large leathery leaves. In some cases stomata are restricted to the abaxial surface, often sunken or enclosed in grooves or depressions and surrounded by hairs, with the effect of limiting water loss by transpiration (Fig. 4.12). Other xeromorphic characters include the presence of a hypodermis or thick epidermis

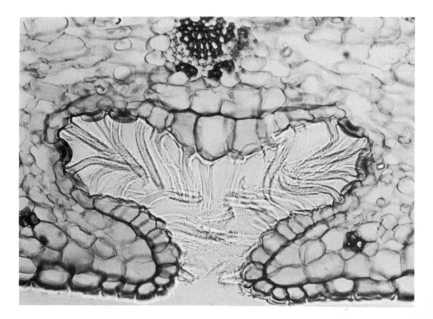

Fig. 4.12 *Erica carnea*. Cross section of leaf, showing abaxial depression lined with unicellular hairs and stomata. (×350)

or cuticle, which diminish the intensity of light reaching photosynthetic tissues, and the presence of large amounts of sclerenchyma, which may provide mechanical support and reduce water loss. Well developed palisade tissue is often correlated with high light intensity, although some xeromorphic leaves also have large air spaces in the mesophyll.

Similarly, certain anatomical features, such as poorly developed sclerenchyma and large air spaces in the ground tissue, are often associated with water plants (hydrophytes). However the physiological bases for adaptive features in leaves are often assumed rather than proved, particularly in xerophytes. Although many plants from dry environments (xerophytes) do show xeromorphic features, there are also mesomorphic plants (without xeromorphic characters) which can survive or even thrive in similarly dry conditions. Conversely many xeromorphic plants can occur in wet habitats. These apparent anomalies can often be explained by other factors; for example by differences in life cycle, as a mesomorphic plant may survive a period of drought in the form of a seed, and a xeromorphic plant in a wet habitat may be subject to some other nutrient deficiency. However they also emphasize the fact that the presence or absence of xeromorphic characteristics is genetically controlled and consequently often of taxonomic importance.

5

The Flower

Angiosperm flowers display a great variety of forms, but in general they are composed of the same basic elements: the perianth, androecium and gynoecium. The perianth is the outer sterile part, with whorls of sepals (calyx) and petals (corolla); the androecium is made up of stamens, which are the site of microsporogenesis; and the central gynoecium, where megasporogenesis occurs, consists of one or more carpels. Each stamen typically has a stalk (filament) bearing an anther, which holds the pollen. The gynoecium has a lower fertile part, the ovary, and an upper sterile part with one or more stigmas, each usually subtended on a style. In hypogynous flowers the ovary is superior, i.e. above the level of insertion of the stamens and perianth parts; in epigynous flowers the ovary is inferior; and in perigynous flowers the stamens and perianth parts are inserted around the ovary. Floral parts may be variously fused in different types of flowers; for example the carpels are often fused into a syncarpous gynoecium. However, in syncarpous gynoecia the stigmas and styles are often separate, the number of branches usually indicating the number of carpels. In flowers where the carpels are not fused, such as in *Ranunculus* (buttercup), the gynoecium is apocarpous.

5.1 Development

5.1.1 Organogenesis
Studies in floral organogenesis have often revealed considerable variation between taxa (Sattler, 1973). For example, in syncarpous gynoecia the carpels may be congenitally fused so that the gynoecium arises as a single structure, or they may be initiated separately and become fused during development. Gynoecia of many species of angiosperms may therefore be apocarpous at initiation, then later become syncarpous for all or part of their length, and finally the carpels may again fall apart after fertilization and prior to dispersal (Endress *et al.*, 1983).

Another aspect of floral variation which requires an ontogenetic approach is the sequence of initiation of the floral organs. In most angiosperm flowers the organ primordia arise from the outside inwards, in centripetal (acropetal) order. However there are a few taxa in which all or some of the organs

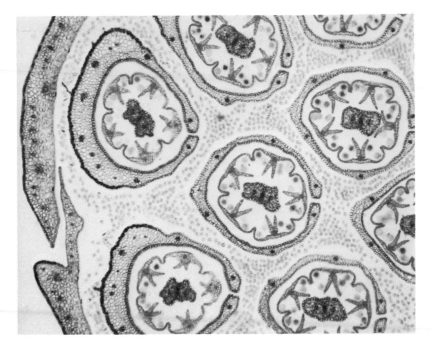

Fig. 5.1 *Taraxacum officinale* (dandelion). Cross section of inflorescence, showing individual florets surrounded by outer petals. (×10)

are initiated in the reverse order, centrifugally, and where this occurs, such as in the stamens of some groups of Palms (Uhl and Moore, 1977), it may be highly characteristic. It is often impossible to determine from mature structure whether stamen initiation is centripetal or centrifugal. In *Drimys* the innermost stamens are the largest and first to dehisce, but their initiation is centripetal (Tucker, 1972).

5.1.2 Microsporogenesis and megasporogenesis

In the developing anther the primary sporogenous cells, which are derived from the same initials as the primary parietal (wall) cells, give rise either directly or by successive mitotic divisions to the microspore mother cells. These in turn each undergo two meiotic divisions, either successive or simultaneous, to form a tetrad of haploid microspores (Maheshwari, 1950). Individual microspores usually separate at this stage, although in some families they remain together as permanent tetrads (Davis, 1966). Prior to release of pollen from the mature anther, the microspores undergo a mitotic division to form a generative cell and a vegetative cell, and in some cases the generative cell further divides in the pollen grain to produce two male gametes, although this latter division more commonly occurs in the pollen tube.

Fig. 5.2 *Coleus barbatus.* Young inflorescence with bracts removed, showing developing flowers. (SEM, ×54)

Maheshwari (1950) summarized the many variations in embryo sac development in angiosperms. As in microsporogenesis, the primary sporogenous or archesporial cells give rise either directly or by successive mitotic divisions to the megaspore mother cell. This then undergoes two meiotic divisions to form four megaspores of varied arrangement and shape. In the majority of angiosperm flowers one megaspore then gives rise to the embryo sac by further mitotic divisions, and the other three megaspores degenerate, although there are many instances where more than one megaspore have a role in embryo sac formation.

5.2   Vasculature

Floral vasculature has been extensively studied in the past, and it has generally been held that the presence of vestigial vascular bundles in certain parts of the flower can be taken to represent vestigial organs, and hence elucidate problems of floral evolution (Puri, 1951). However this theory of vascular conservatism is now controversial, and Schmid (1972) pointed out that the degree of fusion of floral vasculature can be variable even within a single species.

In most flowers the vascular traces to each organ diverge from the central vascular cylinder in a series of whorls, at different levels in the flower. The number of traces to petals and sepals is very varied in different flower types, and the perianth traces often branch dichotomously, as in the leaf. In stamens the most common type of vasculature is a single vascular strand (Fig. 5.3, 5.5), but some families characteristically have three or four stamen

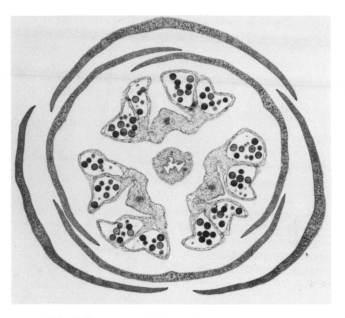

Fig. 5.3 *Crocus sieberi.* Cross section of flower, with six outer tepals (sepals and petals) surrounding three anthers and central style. (×40)

traces, and others, such as Araceae (French, 1986), have diverse and often branching stamen vasculature. The carpellary vascular system is divided into ventral carpellary traces, which diverge into the ovules, and dorsal carpellary traces, which pass up the style into the stigma (Fig. 5.8). The number of vascular bundles in the style of a syncarpous gynoecium is often an indicator of the number of carpels, although sometimes bundles may be branched or fused.

In flowers with very complex vascular systems, such as those of many Magnoliaceae, vasculature is most easily studied in the immature bud, where the ontogenetic sequence of initiation can be observed. Some authors (e.g. Schmid, 1976; Stevenson and Owens, 1978) have suggested that xylem and phloem initiation may be related to functional aspects; for example phloem may develop precociously in areas which require a copious supply of nutrients, such as developing sporogenous tissue.

5.3 Petals and sepals

In their simplest form perianth parts are essentially leaf-like, although there are many flowers with modified and fused petals, and sometimes fused sepals. In cross section both petals and sepals consist of an abaxial and adaxial epidermis enclosing usually three or four or sometimes up to ten layers of undifferentiated isodiametric or elongated cells with many air spaces between them (Fig. 5.3, 5.11). This 'mesophyll' tissue is interspersed with a row of vascular bundles. Petals and sepals are in general less modified than leaves in their internal structure, although they may also contain idioblasts such as crystal-containing cells, or specialized tissues such as a hypodermis. Sepals are frequently green and photosynthetic, and it is common to find stomata and trichomes on surfaces of both petals and sepals.

In insect pollinated flowers the main function of the corolla is to attract insects, and it is consequently the largest and showiest part of the flower. (In wind pollinated flowers the reverse is often the case, the perianth being much reduced or even absent.) Flower colour is mainly controlled by the chemistry of the pigments. Kay *et al.*, (1981), in a survey of sixty plant families, showed that with few exceptions anthocyanins, betalains and ultraviolet absorbing flavonoids are confined to the epidermal cells of the petal, whereas other pigments, such as carotenoids, are found in either the epidermis or the mesophyll. They correlated this information with observations on the surface morphology of the petals. Petal surfaces are either smooth (as in species of *Crocus*), in which case incident light is reflected strongly, or papillate (as in species of *Viola* and *Cistus*) with papillae of various heights, and either one or several per cell (Fig. 5.4). The effect of the domed or papillate cell surface is to guide incident light into the petal, where it is reflected outwards from the inside walls of the epidermal cells or from the multi-faceted walls of mesophyll cells, thus passing through the pigments which are in solution in the cell vacuoles. In some species of *Ranun-*

Fig. 5.4 *Veronica* sp. Petal surface, showing papillate cells with striations. (SEM, ×210)

culus with bright yellow flowers incident light is reflected from starch grains in the subepidermal mesophyll cells.

Many petal surfaces are also strongly striated, which may have the effect of further scattering the incident light into the interior of the petal. Striations on papillae usually run from the base to the apex of the papilla. Gale and Owens (1983) found that in some genera of Commelinaceae petal surfaces were smooth, and in others striated. They also found that cuticular striations occurred on the stamens, staminodes and styles of many species, indicating that these organs also play a part in light absorption patterns and attraction of insects to the flower.

5.4 Androecium

5.4.1 Stamens

Stamen filaments are usually slender and cylindrical, but may be leaf-like or branched. In cross section they have a parenchymatous ground tissue surrounding the usually simple vasculature (section 5.2). The filament surface often bears trichomes, stomata and surface patterning, as in other floral parts.

The anther wall is made up of several layers which are all derived from primary parietal cells, apart from the epidermis, which undergoes only anticlinal divisions during development (Fig. 5.5). The two most distinct layers

Tapetum Pollen Anther locule Vascular bundle Epidermis Endothecium

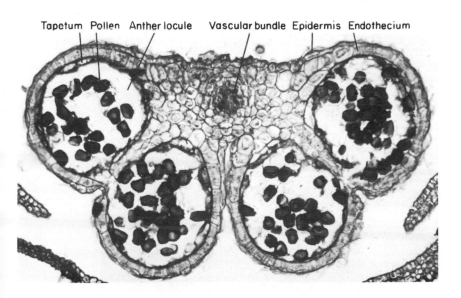

Fig. 5.5 *Sisyrinchium* sp. Cross section of anther. (×150)

are the endothecium, immediately within the epidermis, and the tapetum, which is the innermost layer of cells surrounding the anther locule. Intervening layers usually consist of thin walled cells which are often crushed and destroyed at maturity, although Davis (1966) recorded various types of inner parietal layers. Endothecial cells, most commonly in a single layer, typically develop fibrous wall thickenings which contribute to the anther dehiscence mechanism. Tapetal cells are secretory and full of dense cytoplasm. The contents of the tapetal cells are absorbed by the developing pollen grains in the anther locule, so that by the time the pollen grains are mature the tapetum has usually completely degenerated. In some cases the protoplasts of the tapetal cells break away into the anther locule and coalesce to form a periplasmodium, which also eventually disintegrates.

5.4.2 Pollen

The products of microsporogenesis (section 5.1.2) are the spores or pollen grains (Figs 5.6, 5.7b). These are radially or bilaterally symmetrical bodies with circular or elongated apertures (pores and colpi) through which the pollen tube will emerge, although there are rare examples of non-aperturate pollen grains (Erdtman, 1966). The pollen grain wall has two main layers: the soft inner intine and the hard outer exine, the exine being further subdivided into the outer sculptured sexine and the inner non-sculptured part, the nexine.

Angiosperm pollen is often of taxonomic significance either at or below the family level. Pollen grains vary considerably in size and shape, and there are many different patterns of surface sculpturing, elements of which include granules, warts and spines. In many taxa this surface patterning may

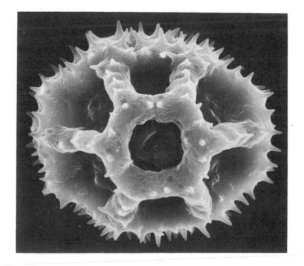

Fig. 5.6 *Taraxacum officinale* (dandelion). Pollen grain. (SEM, ×2000)

be a simple mechanical adaptation, ensuring elasticity of the wall, or providing resistance to collapse. However in a few families with pollen with deeply sculptured or chambered exines, such as Cruciferae, Malvaceae or Compositae, it has been demonstrated that the intercolumellar spaces of the exine carry materials which are derived from the tapetum, and are responsible for control of intraspecific compatibility (Heslop-Harrison, 1976). When the pollen grains are being carried by the pollinator from the anther to the stigma, they dehydrate after contact with the air, and the exine contracts. On the stigmatic surface rehydration occurs, the exine expands and the substances held in both intine and exine are released.

5.5 Gynoecium

5.5.1 Stigma and style
Stigmatic epidermal cells are modified to provide a receptive surface for pollen grains. They are secretory, with a specialized cuticle, and may be slightly domed or with variously elongated papillae. Stigmas can broadly be divided into two types: 'dry', with little or no surface secretions, or 'wet', with surface secretions which assist in pollen germination (Heslop-Harrison and Shivanna, 1977). The stigmatic cuticle usually appears stratified in transverse section, with a lamellated outer layer and reticulate inner layers; and there are many modified types, such as in *Crocus*, which has a chambered cuticle, and *Euphorbia*, where the surface is fenestrated (Heslop-Harrison and Heslop-Harrison, 1982). The cuticle is often torn at maturity as secretions are released. The germinating pollen tubes grow down either between the stigmatic papillae or along the surface of the stylar canal.

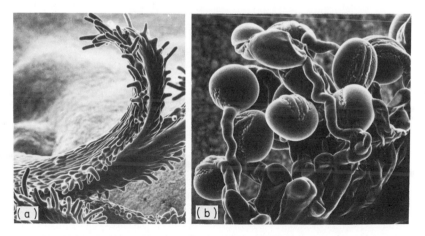

Fig. 5.7 *Anomotheca laxa*. Flower: **(a)** stigmatic surface (×53), **(b)** germinating pollen grains (×230).

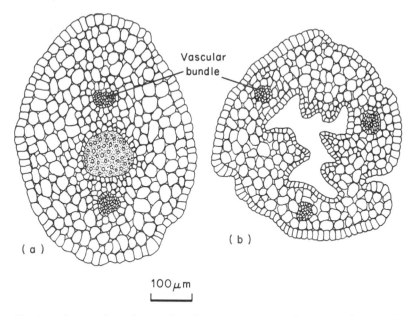

Fig. 5.8 Cross sections of styles of **(a)** *Salvia pratensis* ('closed' style), **(b)** *Crocus speciosus* ('open' style).

The anatomy of the style varies depending on the degree of fusion of the carpels. In many of the higher angiosperms with syncarpous gynoecia the style is solid, with a central specialized secretory tissue, the transmitting tissue, which links the stigma with the centre of the ovary and serves as a nutrient-rich tract for pollen-tube growth (Fig. 5.8a). Other flower types have a central stylar canal lined with secretory, often papillate epidermal cells (Fig. 5.8b), and in a few cases the stigma is borne directly on the top of the ovary. In a comprehensive study of the gynoecium of *Ornithogalum caudatum*, Tilton and Horner (1980) observed that the tissue traversed by the pollen tubes from the tip of the stigma to the base of the locule in the ovary is morphologically similar, and derived from the adaxial epidermis of the carpel or its immediate derivatives.

The ground tissue of the style is parenchymatous, interspersed with a number of vascular bundles.

5.5.2 Ovary

The ovary contains one or more ovules, and each ovule is attached to the ovary wall at a placenta, of which there are usually two per carpel (Fig. 5.9). The term 'placentation' refers to the arrangement of the placentae and the number of locules in the ovary, both of which depend on the degree of fusion of the carpels. In syncarpous ovaries where the carpels are congenitally fused or have grown together postgenitally so that they are fused in the centre of the ovary, there are as many locules as carpels, and placentation is axile.

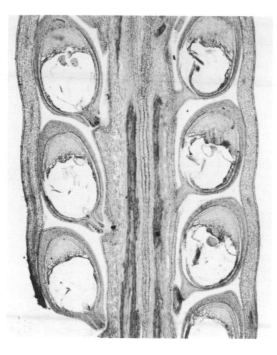

Fig. 5.0 *Crocus sieheri.* Longitudinal section of ovary. (×40)

Where the carpels are fused at their outer edges but not in the centre of the ovary, there is only one locule and placentation is parietal. Other placentation types may occur in unilocular ovaries, for example the placentae may be at the base of the ovary (basal placentation), or on a central column of tissue which is not joined to the ovary wall except at top and bottom (free central placentation).

In species with solid styles, transmitting tissue is continuous from the style through the centre of the ovary, for the growth of pollen tubes to the ovules. There is often an opening (compitum) near the micropyle of each ovule, which enables pollen tubes to grow into an embryo sac. Many taxa have obturators, which are proliferations of secretory tissue around the base of the funiculi, providing nutrients for the developing pollen tubes and guiding them into the micropyles (Tilton and Horner, 1980).

5.5.3 Ovule and embryo sac

Each ovule is attached to the ovary by a funicle, which has a single vascular strand. The ovule wall, which will develop into the seed coat after fertilization (section 6.1), consists of an inner layer, the nucellus, surrounded by one or two (inner and outer) integuments (Fig. 5.10). The region where the nucellus, integuments and funicle connect is termed the chalaza (Fig. 6.1),

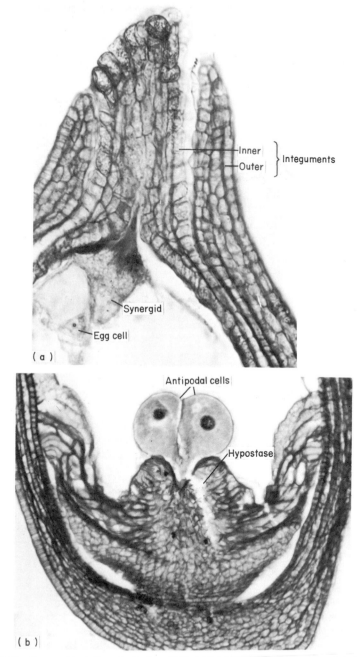

Fig. 5.10 *Crocus sieberi*. Longitudinal sections of ovule at **(a)** micropylar end (×340), **(b)** chalazal end (×135).

often opposite the micropyle, which is a narrow opening at one end of the ovule. Pollen tubes grow through the micropyle during the process of fertilization. Like the tapetum in the anther, the nucellus is usually absent at maturity, having completely degenerated, but at the onset of megasporogenesis it consists either of a single layer of cells or several layers with a well-defined inner and outer epidermis. In some mature ovules the nucellus may have proliferated at the chalazal end of the ovule, opposite the micropyle, to form a hypostase. The hypostase is often refractive and suberinized or lignified, and it may be very large, as in *Crocus*. The vascular bundle in the funicle usually terminates at the chalaza, but in some cases is more extensive.

In many angiosperms the mature embryo sac has eight nuclei, each often

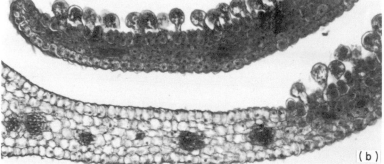

Fig. 5.11 *Tigridia meleagris*. Elaiophores on petal surface. **(a)** SEM (×300), **(b)** Cross section (×130).

enclosed in cell walls. However types with four and sixteen nuclei have also been recorded (Maheshwari, 1950; Davis, 1966). These eight nuclei consist of three antipodal nuclei, usually at the chalazal end, two polar nuclei in the centre, and two synergids and an egg cell at the micropylar end. Many mature embryo sacs have only seven nuclei because the two polar nuclei fuse at an early stage to form a fusion nucleus. In some taxa the antipodal cells degenerate at a very early stage, leaving only five nuclei in the embryo sac. In others, such as *Crocus* (Fig. 5.10), the antipodal cells are very large and persist until after fertilization, often becoming vacuolated. Antipodals may sometimes proliferate to form large numbers of cells (e.g. in grasses). The synergids often have wall thickenings, called the filiform apparatus, which extends into the micropyle.

5.6   Nectaries and elaiophores

Daumann (1970) defined nectaries as clearly localized areas of tissue which regularly secrete nectar, a sugary substance which is attractive to insects. Extrafloral nectaries are found on the vegetative parts, mainly the leaf or petiole, of some plants (e.g. species of *Acacia*). The insects attracted are often ants, which may help to protect the plant from predators. Floral nectaries, on the other hand, usually attract pollinating insects, such as bees or moths, or even more unusual pollinators such as small birds or mammals (Proctor and Yeo, 1973). Nectaries usually consist of secretory epidermal cells with dense cytoplasm, sometimes modified into trichomes. The subepidermal cells may also be secretory, and in some cases nectar passes to the surface through modified stomatal pores. Vascular tissue close to the nectary often consists mainly or entirely of phloem, which transports sugars to the secretory area. Some flowers bear oil-secreting glands, termed elaiophores (Fig. 5.11), which are also attractive to insects, and morphologically often very similar to nectaries (Vogel, 1974).

6

Seeds and fruits

At fertilization, the pollen tube enters the embryo sac through the micropyle and deposits two male nuclei, one of which fuses with the egg nucleus and the other with the two polar nuclei or fusion nucleus. These events initiate the development of the seed.

Seeds are typically composed of three parts: the embryo and endosperm (both products of the double fertilization) and the seed coat, which is formed from the ovule wall (Fig. 6.1). Fruits develop from the ovary or sometimes from the whole flower or inflorescence, and may contain from one to several seeds. The terms 'fruit' and 'seed' are often used loosely; for example the 'seeds' of many Gramineae are actually one-seeded fruits. There are many different types of fruit, depending on how they are derived. They may be dry and dehiscent (e.g. the porous capsule of the poppy, or the legume pod of the pea), dry and indehiscent (e.g. the hazel nut), or fleshy. Fleshy fruits also vary, from drupes (e.g. plums) and berries (e.g. tomatoes), which are both derived from single flowers, to the syconium of the fig, derived from an inflorescence.

The different seed types depend on the relative position and size of the component parts, and the relative position of the hilum, which is the point of attachment of the seed to the placenta. For example, in campylotropous seeds (Fig. 6.1) the ovule has bent over through ninety degrees and become fused to the funicle, which forms the raphe. Variation in seed structure often has an adaptive significance, largely relating to the method of seed dispersal or the ability of the seed to survive dry, cold or otherwise extreme conditions.

6.1   Seed coat

Most angiosperm seeds have a seed coat, or testa, derived from the two integuments or single integument of the ovule (Figs 6.1, 6.2). In bitegmic seeds the term 'testa' is correctly applied only to the outer layer, formed from the outer integument, the part formed from the inner integument being the tegmen. The nucellus often degenerates at an early stage, but may persist, especially at the chalazal end, where it sometimes forms a food storage tissue called the perisperm.

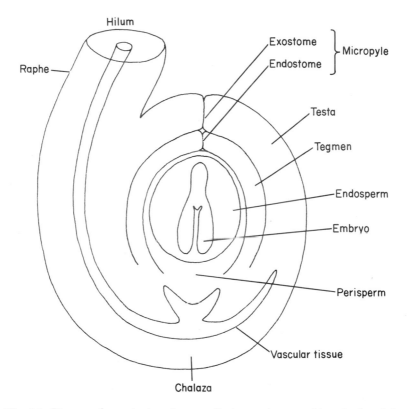

Fig. 6.1 Diagram of organization of a generalized campylotropous bitegmic dicotyledon seed with perisperm. (After Boesewinkel and Boumann, 1984).

Seed coats may be complex multilayered tissues, or simply enlarged ovule walls. They generally include a hard, protective layer formed from all or part of the testa or tegmen. Corner (1976) classified seed coats according to the position of this mechanical layer. For example, in exotestal seed coats the mechanical layer is formed from the outer epidermis of the outer integument, and in endotegmic seed coats it is derived from the inner layer of the inner integument. Sometimes the mechanical layer consists of one or more rows of elongated, palisade-like cells, such as the macrosclereids in the exotesta of many Leguminosae.

Apart from the obvious mechanical protective function, to prevent destruction of the seed by dehydration or predation, the seed coat often has important subsidiary functions, usually related to dispersal, and with corresponding specialized structures (Boesewinkel and Boumann, 1984). For example, seeds may be winged, for wind dispersal, or fleshy, for dispersal by animals. The fleshy part of the seed coat, most commonly formed from part of the outer integument, is termed the sarcotesta, although arils are also

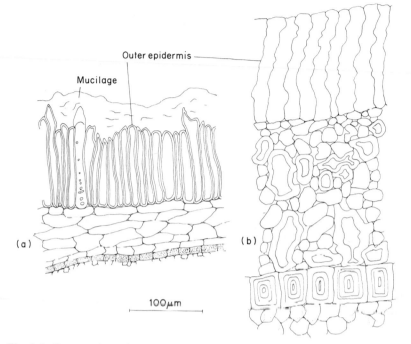

Mucilage

Outer epidermis

(a)

(b)

100μm

Fig. 6.2 Cross sections of testa of **(a)** *Citrus sinensis* (sweet orange) **(b)** *Citrullus vulgaris* (water melon).

fleshy outgrowths of the seed, displaying a great variety of forms. The vasculature of the seed coat usually consists of a single bundle passing from the raphe to the chalaza, but this can vary in extent and degree of branching. In orchid seed coats vasculature is often completely absent.

Seed coat surfaces exhibit a variety of cellular patterns, often with papillate or striate surface sculpturing which may be characteristic of taxa (Fig. 6.3) (Barthlott and Ehler, 1977; Barthlott, 1981). Many seeds have epidermal trichomes, notably *Gossypium* (cotton), where the epidermal outgrowths are an important source of textile fibres. In mucilage seeds, of which there are many examples among angiosperms (e.g. species of *Coleus*), the epidermal cells of the seed coat absorb water and rupture, producing large amounts of slime interspersed with coiled thread-like protuberances. The main purpose of this mucilage production is unknown, but it may be simply to adhere the seed firmly to the soil surface, or to the surface of an animal for dispersal.

6.2 Fruit wall

The fruit wall is called the pericarp if the fruit develops from a single ovary and the fruit wall from the ovary wall. It displays a similar range of variation

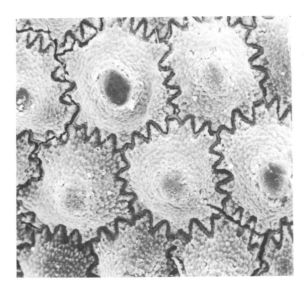

Fig. 6.3 *Silene nutans*. Seed surface; papillate epidermal cells with sinuous anticlinal walls. (SEM, ×260)

to the seed coat, depending particularly on whether the fruit is dry or fleshy. The pericarp is typically divided into three layers: the outer exocarp, central mesocarp, and inner endocarp, although sometimes the three layers are not readily distinguishable. As with the seed coat, at least one layer of the fruit wall often consists of thick walled lignified cells, although in some fleshy fruits (berries), such as the grape, the entire pericarp may consist of thin walled succulent cells. In other fleshy fruits (drupes), such as the peach, the endocarp cells are thick walled, and only the mesocarp is fleshy, the exocarp being a narrow epidermal layer. In the olive the fleshy mesocarp is interspersed with thick walled sclereids (Fig. 6.4).

6.3 Endosperm

The endosperm is usually triploid, formed by fusion of one male nucleus with two female polar nuclei. It is a nutritive, food storage tissue, present in most angiosperm seeds but in greatly varying amounts, and occasionally absent (e.g. in Orchidaceae). There are three main types of endosperm: nuclear, cellular and helobial, defined by differences in early development, although the distinction between the three types is not always clear. In nuclear endosperm the early cell divisions are unaccompanied by cytokinesis (cell wall formation), and the nuclei are initially free in the cytoplasm of the embryo sac, usually surrounding a central vacuole. In most cases of this type endosperm wall formation occurs eventually, but sometimes the

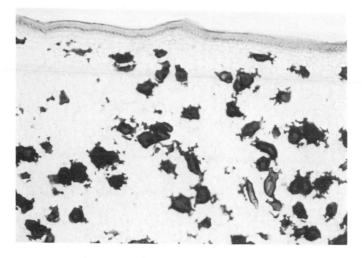

Fig. 6.4 *Olea* sp. (olive). Cross section of pericarp, with numerous branching sclereids. (×60)

nuclei at the chalazal end remain free. In some instances the endosperm develops outgrowths (haustoria) which invade surrounding tissues (Maheshwari, 1950; Vijayaraghavan and Prabhakar, 1984). Coconut palms (*Cocos nucifera*) have nuclear endosperm formation, and the liquid 'coconut milk' contains many free endosperm nuclei.

In the cellular type of endosperm formation, cytokinesis occurs even with the initial cell divisions, although the pattern of cell wall formation varies considerably between different families, and again there are many remarkable instances of development of haustoria. Helobial endosperm formation occurs only in monocotyledons and is considered intermediate between the former two types. The division of the primary endosperm nucleus is accompanied by the formation of a small chalazal chamber and a large micropylar chamber. The nucleus of the micropylar chamber migrates to the top of the embryo sac, and its initial divisions are unaccompanied by cytokinesis, although with later divisions cell walls are formed. The chalazal chamber has far fewer nuclear divisions and its nuclei remain free in the cytoplasm. Vijayaraghavan and Prabhakar (1978) recognized several types of helobial endosperm formation.

Mature endosperm typically consists of tightly-packed cells with either thick or thin walls, and containing food reserve materials such as starch grains or protein bodies, which may be in various characteristic forms.

6.4 Embryo

In a normal angiosperm reproductive system the embryo develops from the diploid fertilized egg cell. Following fertilization the zygote undergoes a

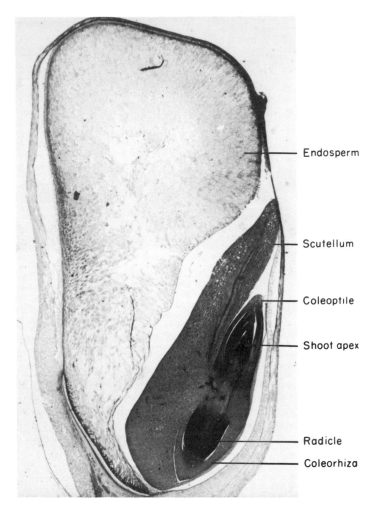

Endosperm

Scutellum

Coleoptile

Shoot apex

Radicle

Coleorhiza

Fig. 6.5   *Zea mays* (maize). Longitudinal section of seed. (×12).

change in volume, either shrinkage or enlargement, before cell division commences, usually initially by a transverse division to form apical and basal cells. The pattern of subsequent cell divisions varies between taxa and has been classified into several types (Maheshwari, 1950). In many cases considerable taxonomic significance is attached to the type of early embryogenesis, and there are distinct differences between the early embryogenesis of dicotyledons and monocotyledons. However most embryos eventually differentiate into a globular mass of cells attached to the wall of the embryo sac by a stalk (the suspensor). The suspensor also exhibits great diversity, and sometimes has a secretory function. It may be uniseriate or multiseriate, but ultimately degenerates.

The main body of the proembryo eventually becomes organized into a structure with root and shoot apices at opposite ends of an axis (the hypocotyl). Many dicotyledon embryos become bilobed as the two cotyledons differentiate (Fig. 6.1) and monocotyledon embryos develop a single, often elongated cotyledon. The degree of differentiation varies considerably in mature embryos; in orchids the embryo remains a simple undifferentiated mass of cells. Some highly differentiated embryos may have, in addition to the hypocotyl and cotyledons, a short primordial root (radicle), often with a root cap, and a shoot bud or short shoot (the epicotyl) developed beyond the cotyledons. Grass embryos (Fig. 6.5) are unique highly differentiated structures with a long specialized cotyledon (the scutellum), an epicotyl surrounded by a sheath (the coleoptile), and a well differentiated radicle also surrounded by a sheath (the coleorhiza). Some grass embryos also have an outgrowth opposite the scutellum, the epiblast, which has been variously interpreted as a second cotyledon or an outgrowth of the first cotyledon or coleorhiza (Natesh and Rau, 1984).

6.5 Seedling

At germination the testa is ruptured and the radicle emerges through the micropyle. The seedling is the most juvenile stage of the plant, immediately

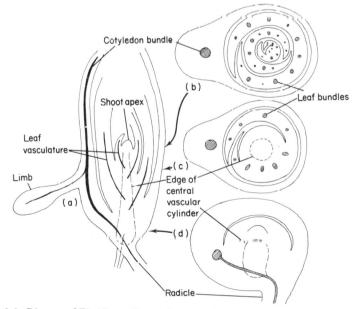

Fig. 6.6 Diagram of *Tigridia* seedling to show vasculature. (a) Longitudinal section. (b) (c) and (d) Cross sections at positions indicated by arrows.

after germination. Seedlings are fairly diverse in structure, depending on the relative size and position of the component parts. Variation in seedling morphology is often of taxonomic significance. Seedlings have a root (radicle) and a hypocotyl, which bears the cotyledons and plumular bud. The bud produces the stem and leaves, which soon resemble those of the mature plant. The hypocotyl varies in size from a swollen food-storage organ to a very short structure which may be almost non-existent, in which case the radicle extends almost to the cotyledonary node. The cotyledons, or seed leaves, usually differ from the first foliage leaves. In some cases, mainly in larger-seeded dicotyledons such as the legume pea or bean (e.g. *Vicia faba*), the cotyledons are fleshy and swollen, with a food-storage function, and remain permanently enclosed in the testa even after the radicle has emerged. This is termed hypogeal germination, as opposed to the more common epigeal germination, where the cotyledons are borne above ground, and are usually photosynthetic. In monocotyledon seedlings the cotyledon typically consists of three parts: a basal sheath, a ligule or ligular sheath, and a limb (Arber, 1925), each part being more or less well-developed. For example, in *Tigridia* seedlings (Fig. 6.6) both the hypocotyl and the basal sheath are extremely reduced and almost non-existent. Cotyledon vasculature is usually much simpler than that of foliage leaves, and may consist of a single vascular bundle, as in *Tigridia*.

Further general reading

Corner, E.J.H. (1976). *The seeds of Dicotyledons.* Cambridge University Press, Cambridge.

Esau, K. (1965). *Plant anatomy.* 2nd ed. John Wiley and Sons Inc. New York, London, Sydney.

Fahn, A. (1979). *Secretory tissues in plants.* Academic Press, London.

Fahn, A. (1982). *Plant anatomy.* 3rd ed. Pergamon Press, Oxford.

Jane, F.W. (1970). *The structure of wood.* 2nd ed. Adam and Charles Black, London.

Juniper, B.E. and Jeffree, C.E. (1983). *Plant surfaces.* Edward Arnold, London.

Maheshwari, P. (1950). *An introduction to the embryology of angiosperms.* McGraw-Hill Book Company, New York, Toronto, London.

Metcalfe, C.R. and Chalk, L. (1979 and 1983). *Anatomy of the dicotyledons,* 2nd ed., Vols I and II. Clarendon Press, Oxford.

References

Arber, A. (1925). *Monocotyledons: a morphological study.* Cambridge University Press, Cambridge.

Arber, A. (1950). *The natural philosophy of plant form.* Cambridge University Press, Cambridge.

Barlow, P.W. (1975). The root cap. In: *The development and form of roots.* J.G. Torrey and D.T. Clarkson (Eds) pp. 21–54. Academic Press, New York and London.

Barthlott, W. and Ehler, N. (1977). Raster-Elektronmikroskopie der Epidermis-Oberflächen von Spermatophyten. *Tropische und Subtropische Pflanzenwelt,* **19**, 1–110.

Barthlott, W. and Wollenweber, E. (1981). Zur Feinstruktur, Chemie und taxonomischen Signifikanz epicuticularer Wachse und ahnlicher Secrete. *Tropische und Subtropische Pflanzenwelt,* **32**, 1–67.

Barthlott, W. (1981). Epidermal and seed surface characters of plants: systematic applicability and some evolutionary aspects. *Nordic Journal of Botany,* **1**, 345–55.

Behnke, H.-D. (1972). Sieve tube plastids in relation to angiosperm systematics – an attempt toward classification by ultrastructural analysis. *Botanical Review,* **38**, 155–97.

Behnke, H.-D. (1975). The basis of Angiosperm phylogeny: ultrastructure. *Annals of the Missouri Botanical Garden,* **62**, 647–63.

Behnke, H.-D. (1981). Sieve element characters. *Nordic Journal of Botany,* **1**, 381–400.

Böcher, T.W. and Lyshede, O.B. (1968). Anatomical studies on xerophytic apophyllous plants. I. *Monttea aphylla, Bulnesia retama* and *Bredemeyera colletioides. Biologiske Skrifter,* **16**, 1–44.

Böcher, T.W. and Lyshede, O.B. (1972). Anatomical studies on xerophytic apophyllous plants. II. Additional species from South American shrub steppes. *Biologiske Skrifter,* **18**, 1–137.

Boesewinkel, F.D. and Boumann, F. (1984). The seed: structure. In *Embryology of Angiosperms.* B.M. Johri (Ed.). pp. 567–608. Springer-Verlag, Berlin.

Brandham, P.E. and Cutler, D.F. (1978). Influence of chromosome variation on the organization of the leaf epidermis in a hybrid *Aloë* (Liliaceae). *Botanical Journal of the Linnean Society,* **77**, 1–16.

Burgess, J. (1985). *An introduction to plant cell development.* Cambridge University Press, Cambridge.

Clowes, F.A.L. (1961). *Apical meristems.* Blackwell, Oxford.

Corner, E.J.H. (1976). *The seeds of Dicotyledons.* Cambridge University Press, Cambridge.

Cutler, D.F. and Brandham, P.E. (1977). Experimental evidence for the genetic control of leaf surface characters in hybrid Aloineae (Liliaceae). *Kew Bulletin*, **32**, 23–42.

Daumann, E. (1970). Das Blutennektarium der Monocotyledon unter besonderer Berucksichtigung seiner systematischen und phylogenetischen Bedeuteng. *Feddes Repertorium*, **80**, 463–590.

Davis, G.L. (1966). *Systematic embryology of the Angiosperms*. Wiley, New York.

DeMason, D.A. (1983). The primary thickening meristem: definition and function in Monocotyledons. *American Journal of Botany*, **70**, 955–62.

Diggle, P.K. and DeMason, D.A (1983). The relationship between the primary thickening meristem and the secondary thickening meristem in *Yucca whipplei* Torr. I. Histology of the mature vegetative stem. *American Journal of Botany*, **70**, 1195–204.

Eames, A.J. and MacDaniels, L.H. (1925). *An introduction to plant anatomy*. McGraw-Hill Book Company, New York.

Endress, P.K., Jenny, M. and Fallen, M.E. (1983). Convergent elaboration of apocarpous gynoecia in higher advanced Dicotyledons (Sapindales, Malvales, Gentianales). *Nordic Journal of Botany*, **3**, 293–300.

Erdtmann, G. (1966). *Pollen morphology and plant taxonomy*. Hafner Publishing Company, New York and London.

Esau, K. (1965). *Plant anatomy*. 2nd ed. John Wiley & Sons, New York.

Fahn, A. (1979). *Secretory tissues in plants*. Academic Press, London.

Fahn, A. (1982). *Plant anatomy*. 3rd ed. Pergamon Press, Oxford.

Feldman, L.J. (1984). The development and dynamics of the root apical meristem. *American Journal of Botany*, **71**, 1308–14.

Fisher, J.B. and French, J.C. (1976). The occurrence of intercalary and uninterrupted meristems in the internodes of tropical Monocotyledons. *American Journal of Botany*, **63**, 510–25.

Fisher, J.B. and French, J.C. (1978). Internodal meristems of Monocotyledons: further studies and a general taxonomic summary. *Annals of Botany*, **42**, 41–50.

Foster, A.S. (1947). Structure and ontogeny of the terminal sclereids in the leaf of *Mouriria huberi* Cogn. *American Journal of Botany*, **34**, 501–14.

Foster, A.S. (1956). Plant idioblasts; remarkable examples of cell specialisation. *Protoplasma*, **46**, 184–93.

Franceschi, V.R. and Horner, H.T. (1980). Calcium oxalate crystals in plants. *Botanical Review*, **46**, 361–427.

French, J.C. (1986). Patterns of stamen Vasculature in the Araceae. *American Journal of Botany*, **73**, 434–49.

Gale, R.M.O. and Owens, S.J. (1983). Cell distribution and surface morphology in petals, androecia and styles of Commelinaceae. *Botanical Journal of the Linnean Society*, **87**, 247–62.

Gifford, E.M. and Corson, G.E. (1971). The shoot apex in seed plants. *Botanical Review*, **37**, 147–229.

Heslop-Harrison, J. (1976). The adaptive significance of the exine. In: *The evolutionary significance of the exine*. I.K. Ferguson and J. Muller (Eds). Academic Press, London.

Heslop-Harrison, J. and Heslop-Harrison, Y. (1982). The specialised cuticles of the receptive surfaces of angiosperm stigmas. In: *The plant cuticle*. D.F. Cutler, K.L. Alvin and C.E. Price (Eds). Academic Press, London.

Heslop-Harrison, J. and Shivanna, K.R. (1977). The receptive surface of the angiosperm stigma. *Annals of Botany*, **41**, 1233–58.

Hickey, L.J. (1973). A classification of the architecture of dicotyledonous leaves. *American Journal of Botany*, **60**, 17–33.

Hickey, L.J. (1979). A revised classification of the architecture of dicotyledonous leaves. In: *Anatomy of the Dicotyledons*, by C.R. Metcalfe and L. Chalk. pp. 25–39. 2nd ed. Vol. 1. Clarendon Press, Oxford.

Howard, R.A. (1974). The stem-node-leaf continuum of the Dicotyledonae. *Journal of the Arnold Arboretum*, **55**, 125–81.

Howard, R.A. (1979). The stem-node-leaf continuum of the Dicotyledonae. In: *Anatomy of the Dicotyledons*, by C.R. Metcalfe and L. Chalk. pp. 76–86. 2nd ed. Vol. 1. Clarendon Press, Oxford.

Jane, F.W. (1970). *The structure of wood*. 2nd ed. Adam and Charles Black, London.

Jernstedt, J.A. (1984). Root contraction in hyacinth. I. Effects of IAA on differential cell expansion. *American Journal of Botany*, **71**, 1080–9.

Juniper B.E. and Jeffree, C.E. (1983). *Plant surfaces*. Edward Arnold, London.

Kaplan, D.R. (1970). Comparative foliar histogenesis in *Acorus calamus* and its bearing on the phyllode theory of monocotyledonous leaves. *American Journal of Botany*, **57**, 331–61.

Kaplan, D.R. (1973a). The Monocotyledons: their evolution and comparative biology. VII. The problem of leaf morphology and evolution in the Monocotyledons. *Quartely Review of Biology*, **48**, 437–51.

Kaplan, D.R. (1973b). Comparative developmental analysis of the heteroblastic leaf series of axillary shoots of *Acorus calamus* L. (Araceae). *Cellule*, **69**, 251–90.

Kaplan, D.R. (1975). Comparative developmental evaluation of the morphology of unifacial leaves in the Monocotyledons. *Botanische Jahrbücher*, **95**, 1–105.

Kaplan, D.R. (1984). Alternative modes of organogenesis in higher plants. In: *Contemporary problems in plant anatomy*. R.A. White and W.C. Dickinson (Eds). pp. 261–300. Academic Press, New York and London.

Kay, Q.O.N., Daoud, H.S. and Stirton. C.H. (1981). Pigment distribution, light reflection and cell structure in petals. *Botanical Journal of the Linnean Society*, **83**, 57–84.

Kuijt, J. (1969). *The biology of Parasitic flowering plants*. University of California Press, Berkeley and Los Angeles.

Maheshwari, P. (1950). *An introduction to the embryology of the Angiosperms*. McGraw-Hill Book Company, New York.

McCully, M.E. (1975). The development of lateral roots. In: *The development and function of roots*. J.G. Torrey and D.T. Clarkson (Eds). pp. 105–24. Academic Press, London.

Metcalfe, C.R. and Chalk, L. (1979 and 1983). *Anatomy of the Dicotyledons*. 2nd ed. Vols. I and II. Clarendon Press, Oxford .

Miles, A. (1978). *Photomicrographs of world woods*. HMSO, London.

Natesh, S. and Rao, M.A. (1984). The embryo. In: *Embryology of Angiosperms*. B.M. Johri (Ed.). pp. 377–434. Springer-Verlag, Berlin.

Obaton, M. (1960). Les lianes ligneuses à structure anormale des forêts denses d'Afrique occidentale. *Annales des Sciences Naturelles, Botanique*. ser. 12, **1**, 1–120.

Proctor, M. and Yeo, P. (1973). *The pollination of flowers*. Collins, London.

Puri, V. (1951). The role of floral anatomy in the solution of morphological problems. *Botanical Review*, **17**, 471–553.

Rasmussen, H. (1983). Stomatal development in families of Liliales. *Botanische Jahrbücher*, **104**, 261–87.

Sattler, R. (1973). *Organogenesis of flowers. A photographic atlas.* University of Toronto Press, Toronto.

Schenck, H. (1893). Beiträge zur Anatomie der Lianen. *Schimper's Botanischer Mittheilungen aus den Tropen,* Heft 5.

Schmid, R. (1972). Floral bundle fusion and vascular conservatism. *Taxon,* **21**, 429–46.

Schmid, R. (1976). Filament histology and anther dehiscence. *Botanical Journal of the Linnean Society,* **73**, 303–15.

Schmidt, A. (1924). Histologische Studien an phanerogamen Vegetationspunkten. *Botanisches Archiv,* **8**, 345–404.

Schweingruber, F.H. (1978). *Microscopic wood anatomy.* Swiss Federal Institute of Forestry research, Birmensdorf.

Shigo, A.L. (1985). How tree branches are attached to trunks. *Canadian Journal of Botany,* **63**, 1391–401.

Stevenson, D.W. and Owens, S.J. (1978). Some aspects of the reproductive morphology of *Gibasis venustula* (Kunth) D.R. Hunt (Commelinaceae). *Botanical Journal of the Linnean Society,* **77**, 157–75.

Stevenson, D.W. (1980). Radial growth in *Beaucarnea recurvata. American Journal of Botany,* **67**, 476–489.

Stevenson, D.W. and Fisher, J.B. (1980). The developmental relationship between primary and secondary thickening growth in *Cordyline* (Agavaceae). *Botanical Gazette,* **14**, 264–8.

Tilton, V.R. and Horner, H.T. (1980). Stigma, style and obturator of *Ornithogalum caudatum* (Liliaceae) and their function in the reproductive process. *American Journal of Botany.* **67**, 1113–31.

Tomlinson, P.B. (1974). Development of the stomatal complex as a taxonomic character in Monocotyledons. *Taxon,* **23**, 109–28.

Tomlinson, P.B. and Zimmerman, M.H. (1969). Vascular anatomy of Monocotyledons with secondary growth – an introduction. *Journal of the Arnold Arboretum,* **50**, 159–79.

Tschirch, A. (1889). *Angewandte Pflanzenanatomie.* Wien and Leipzig.

Tucker, S.C. (1972). The role of ontogenetic evidence in floral morphology. In: *Advances in plant morphology.* Y.F. Murty *et al.* (Eds). Sarita Prakashan, Meerut, India.

Uhl, N.W. and Moore, H.E. (1977). Centrifugal stamen initiation in phytelephantoid palms. *American Journal of Botany,* **64**, 1152–61.

Vijayaraghavan, M.R. and Prabhakar, K. (1984). The endosperm. In: *Embryology of Angiosperms.* B.M. Johri (Ed.). pp. 319–370. Springer-Verlag, Berlin.

Vogel, S. (1974). Olbumen und olsammelnde Bienen. *Tropische und Subtropische Pflanzenwelt,* **7**, 285–577.

Wilkinson, H.P. (1979). The plant surface. In: *Anatomy of the Dicotyledons,* by C.R. Metcalfe and L. Chalk. pp. 97–165. Clarendon Press, Oxford.

Zimmerman, M.H. and Tomlinson, P.B. (1965). Anatomy of the palm *Rhapis excelsa.* I. Mature vegetative axis. *Journal of the Arnold Arboretum,* **46**, 160–80.

Zimmerman, M.H. and Tomlinson, P.B. (1966). Analysis of complex vascular systems in plants: optical shuttle method. *Science,* **152**, 122–42.

Zimmerman, M.H. and Tomlinson, P.B. (1972). The vascular system of monocotyledonous stems. *Botanical Gazette,* **133**, 141–55.

Index

Page numbers in bold indicate reference to an illustration.